Flexible Innovation
Technological Alliances in Canadian Industry

Technological alliances have begun to change the way high-technology companies conduct research and development. While most companies still carry out R&D in isolation, the secretive, closed laboratory is becoming a thing of the past. *Flexible Innovation*, the first study of these changes within the Canadian context, is a comprehensive look at the state of technical collaboration in high-technology firms and a guide-book for formulating and facilitating technical alliances.

Basing his study on in-depth interviews in more than 130 companies across Canada, Jorge Niosi analyses the scope of collaborative research activities – both domestic and international – in the fields of bio-technology, electronics, advanced materials, and manufacturing of transportation equipment. He describes successful patterns of collab-oration, obstacles and limitations, and the role of public policy, universities, and government laboratories in technological alliances and compares Canadian partnerships and public policy with similar patterns in Europe, the United States, and Japan.

JORGE NIOSI is professor of administrative science and director of the Centre for Interuniversity Research on Science and Technology, Université du Québec à Montréal.

Flexible Innovation

*Technological Alliances in
Canadian Industry*

JORGE NIOSI

with the collaboration of
MARYSE BERGERON, MICHÈLE SAWCHUCK,
and NATHALIE HADE

McGill-Queen's University Press
Montreal & Kingston • London • Buffalo

Legal deposit fourth quarter 1995
Bibliothèque nationale du Québec

Printed in Canada on acid-free paper

This book has been published with the help of a grant
from the Social Science Federation of Canada, using
funds provided by the Social Sciences and Humanities
Research Council of Canada.

McGill-Queen's University Press is grateful to the Canada
Council for support of its publishing program.

Canadian Cataloguing in Publication Data

Niosi, Jorge, 1945–
   Flexible innovation : technological alliances in Canadian
   industry

   Includes bibliographical references and index.
   ISBN 0-7735-1334-5 (bound) –
   ISBN 0-7735-1335-3 (pbk.)

   1. Cooperative industrial research – Canada 2. Research,
   Industrial – Canada 3. High technology industries –
   Canada. I. Bergeron, Maryse II. Hade, Nathalie
   III. Sawchuck, Michèle, 1965–   IV. Title.

   T177.C3N56 1995   607'.271   C95-900675-3

Typeset in New Baskerville 10/12
by Caractéra production graphique, Quebec City

Contents

Tables   vii

Preface   xi

1   Theories of Technological Alliances   3

2   A Cross-Country Comparison of Government Policy   25

3   Electronics   38
    (with the collaboration of Maryse Bergeron)

4   Advanced Materials   59

5   Biotechnology   77
    (with the collaboration of Nathalie Hade)

6   Transportation Equipment   99
    (with the collaboration of Michèle Sawchuck)

7   Implications for Government and Business   113

8   Conclusions   122

Bibliography   131

Index   141

Tables

1 Strategic European alliances by sector and rationale   12

2 Technical cooperation   24

3 Collaborations by electronics firms   42

4 Number of partners in electronics alliances   43

5 Managing the electronics alliance   44

6 Solving the intellectual-property dilemma in electronics alliances   44

7 Explaining the choice of partner in electronics   45

8 Reasons for technical alliances in electronics   46

9 Advantages of technical alliances in electronics   46

10 Assessing the results of R&D alliances in electronics   46

11 Firm size and number of collaborations in electronics   47

12 Firm size and R&D collaborations as a percentage of total R&D expenditures in electronics   48

13 Size of firms and average contribution to alliances in electronics   48

14 Goals of alliances in electronics, by industry   49

15 Legal arrangements for collaboration in electronics, by industry   49

16 Sources of funding and difficulties during the electronics alliance   50

17 Geographic distribution of the partners in electronics alliances   50

18 Differences in international and regional alliances in electronics   51

19 Differences in electronics alliance goals, by industry   52

20 Reasons for choice of partners in electronics, by industry   53

21 Differences in electronics agreements, by industry   53

22 Number of collaborations by advanced materials firms   65

23 Number of partners in advanced materials alliances   65

24 Managing the advanced materials alliance   65

25 Choosing the advanced materials partner   65

26 Explaining technical alliances in advanced materials   66

27 Advantages drawn from advanced materials alliances   67

28 Assessing the results of R&D alliances in advanced materials   68

29 Size of firms and R&D effort in advanced materials   68

30 Size of firms and collaborative R&D in advanced materials   69

31 Size of firms and management of the advanced materials alliance   69

32 Size of firms and ownership of R&D results in advanced materials alliances   69

33 Public financing and difficulties in the negotiation of advanced materials alliances   70

34 Scope of advanced materials alliances   70

35 International versus regional advanced materials alliances   71

36   Industrial breakdown of advanced materials
     alliances   71

37   Number of collaborations per firm in biotechnology   85

38   Number of partners in biotechnology alliances   85

39   Managing the biotechnology alliance   86

40   Explaining the choice of biotechnology partners   86

41   Reasons for alliances in biotechnology   87

42   Advantages of biotechnology alliances   88

43   Assessing the results of R&D alliances in
     biotechnology   88

44   Size of biotechnology firms and R&D effort   89

45   Size of biotechnology firms and R&D personnel   89

46   Age and size of biotechnology firms   90

47   Age and products of biotechnology firms   90

48   Age and sales of biotechnology firms   90

49   Age of biotechnology firms and R&D effort   91

50   Age of biotechnology firms and age of alliances   91

51   Difficulties in biotechnology alliances and source of
     funding   91

52   Geographic scope of biotechnology alliances   92

53   Membership of international and regional biotechnology
     alliances   93

54   Age of biotechnology firms, by industry   93

55   Differences in biotechnology budgets, by industry   94

56   Biotechnology collaborations, by industry   94

57   Biotechnology patenting, by industry   94

58   Number of transportation equipment alliances, per
     firm   103

59   Number of partners in transportation equipment
     alliances   103

60   Managing the transportation equipment alliance   103

61 Solving the intellectual-property dilemma for transportation equipment 104

62 Explaining the choice of partners in transportation equipment alliances 105

63 Public financing in transportation equipment alliances 105

64 Reasons for transportation equipment alliances 105

65 Advantages of transportation equipment alliances 106

66 Assessing the results of R&D in transportation equipment alliances 107

67 R&D effort relative to transportation equipment sales 107

68 Cooperative R&D effort in transportation equipment relative to total R&D expenditures 107

69 Transportation equipment alliances according to the geographic distribution of the majority of the partners 109

70 Membership of international and regional transportation equipment alliances 109

71 Joint patenting in Canada, 1980–1989 120

72 Modes of cooperation in technological alliances 124

Preface

During the 1980s, strategic alliances became one of the most important organizational innovations in modern business; they linked companies, universities, public laboratories, and other economic agencies through financial, technical, or commercial networks. Technological alliances are at the centre of the present collaborative trend. They are revolutionizing the way corporations innovate, conduct research and development (R&D), and trade technology. While most companies still do R&D in isolation, the secretive, closed laboratory is losing its appeal. Few corporations are able to go it alone, because the costs, speed, uncertainties, and risks of technological change have increased dramatically. Technical cooperation is not a passing fashion but the new way of doing research during the present triple technological revolution in electronics, biotechnology, and advanced materials. Few companies, not even the largest, can now master the multifarious dimensions of a major technological research project.

Because the absorption of rivals is often too costly, in the permanent drive for corporate diversification, mergers and acquisitions are less common, and technical cooperation has increased. Cooperation is much more flexible, less expensive, and less risky than outright acquisition of rivals and suppliers, and conglomerates are less efficient than specialized firms conducting strategic alliances.

From a theoretical as well as an empirical point of view, technical cooperation is largely unexplored. In Canada, governments spend hundreds of millions of dollars in support of collaborative research. However, there has been little empirical analysis and evaluation of the characteristics, benefits, and costs of these organizational novelties. Technological alliances have been completely neglected in conventional neoclassical economics, and they can be only weakly linked to industrial, evolutionary, and institutional economics and to management theory.

Because existing theories provide an incomplete framework for analysing the dynamics and benefits of strategic alliances, I decided to study alliances using an inductive methodology. I began, quite properly, with a preliminary examination of the existing literature, both empirical and theoretical, on technological collaboration and alliances. I took into account the conceptual and comparative forerunners, written almost without exception by European and American social scientists, who tried to make sense of alliances. I found that most Canadian high-technology firms did in fact belong to alliances, and I constructed survey samples in the sectors where most of the alliances were taking place. The major themes selected for the survey are of interest both to academics and practitioners: organizational arrangements, intellectual-property issues, the evolution of partnerships over time, government initiatives and funding. I also tried to identify the advantages that corporations sought in alliances and the benefits they had actually drawn from collaborative research and development.

Nearly 130 in-depth interviews were conducted across Canada. The results tended to disconfirm some existing hypotheses, to confirm others, and to pave the way for the development of a new theory. I believe the survey results provide us with first-class, first-hand material that gives a broad picture of technological collaboration, both domestic and international, in Canadian high-technology firms. The response of individuals firms in our interviews remain confidential, but the results are generalized in the statistical tables in this book. Only public information was used to illustrate particular examples of Canadian alliances; the source of this information is indicated in the references.

I begin in chapter 1 by exploring the theoretical underpinnings of economic cooperation and by sketching a new theory of collective innovation. In chapter 2, I present a cross-country comparison of government policy on technological alliances. Chapters 3 to 6 then turn to the analysis of the evidence from the four industries or technologies in which Canadian alliances are concentrated: electronics, advanced materials, biotechnology, and transportation equipment. I consider the implications of alliances for both governments and business in chapter 7, and offer a brief conclusion in chapter 8 that underlines the main findings and revisits the initial theoretical discussion, in the light of the empirical results. I also suggest some new themes for the social sciences research agenda.

This study could not have been conducted without the help of three of my graduate assistants: Maryse Bergeron, Nathalie Hade, and Michèle Sawchuck. They identified approximately half of the nearly 130 companies interviewed and visited nearly 70 companies across Canada, a painstaking undertaking indeed. They wrote three of the empirical chapters jointly with me. Nadia Marzouki and Jérôme Allaire did the statistical work. Dr John Baldwin (Statistics Canada, Ottawa), Michel Delapierre (Université de Paris-Nanterre, France) and Kevin Fitzgibbons, (National Research Council of Canada, Ottawa) read the manuscript and made useful comments. Odette Dallaire helped with the secretarial work. The Social Science Research Council of Canada, the Fonds pour les chercheurs et l'aide à la recherche (FCAR) in Quebec, and the Economic Council of Canada financed different parts of this research. I remain responsible for any errors or omissions in this book.

Some of the ideas and data included in this report have already been published. Chapter 3, on electronics, appeared in *Technovation* (Niosi and Bergeron 1992) and chapter 4, on advanced materials, in *R&D Management* (Niosi 1993).

Flexible Innovation

1 Theories of Technological Alliances

Technological alliances are phenomena in seach of a theory. No available analytical framework can explain the existence and rapid development of either interfirm cooperation in research and development or technical cooperation between industry and university, public laboratories, or other government agencies. This chapter reviews some of the ideas that the theoretical literature, from different currents in economics and management studies, has developed about technological cooperation and alliances.

A technological alliance is a particular case of interfirm cooperation. In a spectrum ranging from informal knowledge-sharing (von Hippel 1987) to mergers and acquisitions (and thus to a total consolidation of the firms' technical assets), alliances stand somewhere in the middle. They are a median solution between soft and hard technical interaction among business enterprises. More precisely, as defined here, technological alliances are long-term contractual agreements between two or more enterprises aiming at the development of new or improved product or process technologies.[1] (Sometimes, research universities and government laboratories are also associated.) Each of the cooperative firms usually contributes various kinds of resources (for example, human, technological, or financial resources, capital equipment, or laboratories) to achieve the common goal and then share the intellectual-property results (Mariti and Smiley 1983; Fusfeld and Haklisch 1985).

---

1 Throughout this book, "technical" or "technological alliances," "partnerships" and "collaborations" are synonymous.

By insisting that they are long-term, our definition of technological alliances puts them in the realm of strategy: our analysis focuses on strategic technological alliances. Short-term, informal technological cooperation, usually involving small amounts of resources and no formal agreements, is thus excluded from this study. Our definitior of alliances also excludes technology transfer, that is, the sale without major alteration of a technology from one economic agent to another.

Technological alliances are organized either through joint ventures or, more often, through written agreements between the partners. An alliance may or may not involve the acquisition, by one partner, of a minority, noncontrolling share-holding position over another. An alliance may be classified as *national* in scope (even if all members belong to one region of a country) or *international* (if members are based in different countries). National and international alliances differ significantly in their goals and management techniques.

Since technological alliances are intermediate forms of organization between hierarchies and markets, they are difficult to classify in received theory. Two major types of alliances appear in the literature: *horizontal* alliances of competing firms operating in the same industry and *vertical*, or client-supplier, alliances. But the complementary assets involved in technical alliances may not be horizontally or vertically linked, since technology is anything but homogeneous, perfectly codified, and well-defined.

I begin this chapter with a review of the problems that technological cooperation poses for existing theory (neoclassical, industrial, regional, and transaction-cost economics), and I then turn to the more empirical, managerial literature on alliances. With the theoretical and empirical problems clearly in mind, I then sketch the fundamentals of a theory of flexible or cooperative innovation that is necessary if we are to understand all forms of technological collaboration, including alliances. The theory of flexible innovation is a major departure from current assumptions in economics and theories of the management of technology, although it does gather together the most useful new approaches and empirical research from both areas.

TECHNOLOGICAL COOPERATION IN ECONOMICS AND MANAGEMENT

Neoclassicism

Technological cooperation among firms, or between firms and universities or public laboratories, is not easy to understand within the framework of existing neoclassical thought. The orthodox neoclassical paradigm is based on a number of assumptions that preclude the analysis of technological cooperation. These assumptions include the following:

1 Firms are profit maximizers with no other motives and behaviours.
2 Firms choose technology within a specific set of options specified by well-defined production functions.
3 Technical knowledge is explicit, codified, public, perfectly divisible, and free.
4 Technological and scientific change at the macro level is a noneconomic phenomenon.
5 Flows of information among firms are reduced to price and output knowledge.
6 Competition, and particularly perfect competition, is the main type of relationship among firms, and it is essentially limited to price competition, since products are homogeneous. Only some neoclassical economists, for example, Hayek (1978), argue that perfect competition provides a framework for technical development and innovation.
7 Economic agents are perfectly rational and possess all the relevant information available, whether it is technical, commercial, or some other information.

In order to explain cooperative organizations started by small and medium-sized enterprises (SMEs) and to explain personal consumption, the neoclassical literature does recognize some special cases of technical collaboration. Marshall (1890) accepted that some cooperatives were useful, but only for organizing collective purchases and sales in agriculture and public utilities. Norris and Vaizey saw research associations as useful instruments for promoting innovation by SMEs, but noted that they were "not suited for industries where one or two firms have a dominant position" (1973, 109).

However, a recent, but growing, empirical literature has shown that the fundamental assumptions of the neoclassical paradigm do not provide a reasonable basis for understanding technological change (Nelson and Winter 1982; Rosenberg 1982; Elster 1983; Dosi, Freeman, et al. 1988). More specifically:

1 Firms do not maximize either profits or sales or any other variable, but rather choose the best option from the specific set of capabilities, rules, and routines they possess and can master.
2 Technological choice is uncertain, risky, and costly, because firms have only imperfect knowledge of the technical options that can be purchased from their environment and of the eventual costs of internally produced technology.

3 Technical knowledge is only partially codified, public, generic, and free. It is also partially tacit, uncodified, and industry- and firm-specific. Technical knowledge necessarily involves important elements of personal, nonverbalized expertise. Moreover, whether it is codified or not, enterprises keep secret a good part of their internally produced knowledge, thereby making technical knowledge even more opaque, less explicit, public, and free.

4 Technological change at the macro level is often the result of the firms' efforts to modify their technical routines, mainly through investments in research and development (R&D); technical change is, thus, internal to the economic system and a key to understanding its dynamics.

5 Technological flows among firms and between firms and other areas of their environment (universities, public laboratories, government) are frequent and, indeed, vital to the enterprises' R&D activities. These flows take many forms and include assistance to technical conferences, subscriptions to publications organized by industry associations, the contracting-out of research projects to universities and public laboratories, and the adoption of public technical standards.

6 Competition is only one of the typical relationships among firms, and although the competition may be pure and perfect, it may also be imperfect and monopolistic. Firms pursue other types of non-competitive relationships when, for example, they imitate another firm's technical or organizational capabilities, when they engage in legal or collusive cooperation for technical, commercial, political, or other purposes, and when they engage in litigation.

7 Bounded rationality is a much more realistic assumption than perfect rationality for the analysis of technical transactions: economic agents are rational, but their knowledge is limited. Thus, we can explain their search for new, complementary knowledge.

A theory of technological alliances and innovation needs these more realistic assumptions: the rationality of economic agents is bounded; the technological environment is opaque, with costly, risky, and uncertain options; there are technical flows among firms; endogenous technological change is determined – at least partially – by the firms' R&D investments; and technical knowledge is a quasi-public good. I shall say more about the last assumption later in this chapter.

Industrial Economics

Industrial economics is more useful than neoclassicism for the study of technical alliances. Joseph Schumpeter (1911, 1942) created a definite link between technology and the basic themes of industrial economics when he proposed, mainly in his later writings, that innovation results from the R&D of large firms operating in oligopolistic markets. While industrial economics is still cast in the neoclassical form, some of the more stringent assumptions of orthodox analysis have been abandoned in favour of new assumptions. The novelties include the following:

1 *The assumption of increasing returns to scale for R&D.* Since efficiency increases with the size of economic units, including R&D units (Scherer 1970), R&D economies of scale provide one explanation for technical cooperation. If large corporations possess the resources necessary to develop a large portfolio of research projects, to purchase expensive specialized equipment, and to hire highly qualified specialists, the same holds true for a group of independent firms organized to conduct R&D. However, this conclusion is only implicit in industrial economics, which has dealt exclusively with R&D economies of scale attained *within* the firm and not with efficiencies achieved through the *cooperation* of *independent* firms.

2 *The assumption of imperfect markets, particularly information markets.* If information markets are imperfect, they are risky, uncertain, and costly. We might deduce that alliances can reduce the risks, uncertainties, and costs of collecting opaque and expensive technical information from different sources. Industrial economics, however, has not gone that far.

3 *The assumption of economies of scope.* If existing facilities allow an industrial corporation to produce several different outputs, "economies of scope are said to exist when it is cheaper to produce the outputs in combination rather than separately" (Baumol and Braunstein 1977). Again, while industrial economics has explored the search for economies of scope by individual firms, it has neglected the collective search for economies of scope. The collective search poses a problem that industrial economics has not even formulated: it is left to the theory of cooperative innovation.

Industrial economics has not explored technical cooperation in economies of scale or scope because, first, its roots are neoclassical: neoclassical theory assumes that information about prices and outputs flows between firms, but not information about technology. Thus, technical cooperation cannot appear as a central problem for industrial economics. Second, industrial economics assumes that imperfect competition is widespread, and considers it the main source of such inefficiencies and unfair practices as predatory behaviour, price discrimination, and output restriction (Blair 1982). Since the norma-tive goals of industrial economics are antitrust, even the most empir-ical industrial economists cannot find any interest in the analysis of interfirm technical cooperation. In fact, a large share of the industrial economics literature on technology has simply centred on Schum-peter's themes: the impact of firm size and market concentration on R&D, patenting, and innovation (Bonin and Desranleau 1988).

Transaction Costs

One important current of economic thought within the management tradition is transaction-cost theory, which excludes cooperation and recognizes only two major types of economic organizations: markets and hierarchies (Coase 1937; Williamson 1975). Firms use the market when transaction costs, defined as the costs of planning, adapting, and monitoring task completion under independent governance struc-tures (Williamson 1985), are lower than the organizational costs of hierarchies. But uncertainty, risk, and market imperfections raise transaction costs and make hierarchies a more efficient choice than markets, especially for technology transactions. Transaction-cost theory has been used primarily for the analysis of production (Pisano 1990). There are few studies of R&D within this framework, mainly because the principal authors in the school argue that R&D is better organized by hierarchies (or internal markets) than by independent units through arm's-length transactions (Williamson 1975, 1985; Teece 1988).

It is not surprising that Lundvall (1990) has challenged the rele-vance of transaction-cost theory for analysing the present-day behav-iour of firms. When R&D alliances can be counted by the thousands and when their benefits seem to overwhelm transaction costs, one may well wonder whether the theory can explain the management of R&D. In the late 1980s and early 1990s, it became clear that firms tend to prefer intermediate forms of organization, such as alliances, to hier-archies. As a result, in a major departure from his previous work, Teece (1989) had to admit that "cooperation is usually necessary to promote competition, particularly when industries are fragmented. Very few firms can 'go it alone' any more. Cooperation in turn requires inter-firm agreements and alliances, [and] the boundaries of the firm can no longer be assessed independently of the cooperative relationships which particular innovating firms may have forged" (3–4). Moreover some critics have identified major conceptual flaws in transaction-cost theory, including vagueness (even in the concept of transaction cost itself) and conflicting and unrealistic assumptions (Kay 1992). On the other hand, Pisano (1990) did find that transaction costs were responsible for the absorption of small biotechnology firms by large pharmaceutical corporations.

Regional Economics

A massive literature in economic geography and regional economics has analysed technological collaborations as "agglomeration effects." High-technology firms concentrate in one geographical area in order to capture externalities from research universities, public laboratories, regional government programs, or regionally based advanced technology firms (Breheny 1988; Markusen, Hall, et al. 1986; Saxenian 1991; Smith 1981; Storper and Harrison 1991). Piore and Sabel (1984) were the first to focus academic attention on the existence of local networks of innovators. On this approach, collaborations are assumed to be mainly regional, urban phenomena involving firms, universities, and public laboratories within the same area. Nevertheless, many authors (among them Mowery 1989; Gomes-Casseres 1992) have shown that international technical alliances are increasingly an element in cross-border economic relations, along with multinational investment and trade. Technological cooperation is by no means limited to industrial districts, or even confined to national borders, but is increasingly an international phenomenon.

TECHNOLOGICAL ALLIANCES IN MANAGEMENT THEORY

Cooperative research traditionally received little attention in the literature on the management of technology. Cooperative R&D was supposed to be of interest only to small industrial firms (Hawthorne 1978). Only in the late 1980s was a more empirical literature, without a clearly defined link to any specific current of thought, developed within management departments in order to explain technical alliances.

Why Do Alliances Exist?

Many factors have been mentioned in the management literature to explain technological alliances and cooperation among independent firms (Chesnais 1988; Dussauge, Garrette, et al. 1988; Hagedoorn 1990; Jorde and Teece 1989). The following are the most frequent:

1 *Realizing R&D economies of scale.* In the pursuit of a common technological goal, companies and other research units may cooperate

either to reduce the R&D costs for each partner or to attain a critical mass of resources permitting them to conduct large-scale projects that would be out of reach for any particular firm. Mergers and acquisitions of competitors may be too costly compared with horizontal R&D agreements. Also, with increasing technological complexity and interconnection, the costs of R&D have been escalating. This hypothesis predicts a larger cooperative effort from smaller firms than from larger ones.

2 *Accelerating innovation.* The organization of companies possessing complementary technical assets (for example, product and process technology, design skills and manufacturing experience) may permit them to achieve faster results than each company would achieve through individual search and learning.

3 *Capturing the knowledge of users.* The acquisition and accumulation of technical assets traditionally takes place internally, within firms, either through R&D (learning by searching) or through manufacturing (learning by doing). However, an increasing number of authors have suggested that users also acquire technical knowledge about products through long-term, repetitive use and that such knowledge is different from knowledge acquired in the laboratory or the plant (von Hippel 1977; Rosenberg 1982; Lundvall 1988). This explanation predicts that alliances will occur between users and producers.

4 *Reducing risk and uncertainty.* In an economic world of technological turbulence, opening markets, and increased competition, the association of several independent firms for R&D projects increases the chances that each partner will succeed and remain in the market.

5 *Coping with short life cycles of products.* Some of the new technologies, particularly electronics, have very short life cycles. In-house R&D may not be sufficient to keep up with the competition when the product or process is not expected to last more than a few years. It may then be worthless to do long-term, in-house R&D, because the companies' technical assets suffer rapid depreciation.

6 *Capturing other complementary assets.* Companies may form alliances because some of them own large or critical technical assets but possess small distribution networks and have limited access to funds, while others have a different range of specialized assets.

7 *Searching for standards.* In the electronics industries, the search for common standards is increasing, since compatibility with complementary products may determine the success or the failure of, for example, computers, telecommunications equipment, software, numerically controlled machines, and various consumer products.

The development of standards may require collective research from groups of firms.

8 *Using new methods of management.* Traditional American methods of management emphasized arm's-length transactions between assemblers and suppliers. According to Porter (1980), large assemblers would be better served by organizing competition among their many suppliers. The Japanese method of managing assembler-supplier relations is just the opposite of the traditional American method: assemblers organize long-term relationships (including R&D cooperation) with their suppliers in order to ensure total quality, reliability, and high performance of parts and components.

9 *Responding to government incentives.* Governments have been sharing the costs of collaborative R&D in order to increase the competitiveness of domestic companies, accelerate technological diffusion, nurture the development of new industries, reduce duplication of efforts, and foster interactive learning. Firms may collaborate to benefit from government support for R&D.

10 *Capturing regional externalities.* High-technology firms may collaborate with other firms, universities, and public laboratories in order to capture externalities from local R&D.

Table 1, which reports the results of a European survey by the *Economist*, summarizes the respondents' reasons for creating strategic technological alliances of the relative importance of several items in our list.

The Management of Technological Alliances

The management of external cooperation is completely different from the management of innovation within the firm. Within the firm, the key issues are the choice between one central or several specialized R&D laboratories, the interface between the R&D department and the other functions (manufacturing, marketing, finance, and the like), the hiring of the appropriate research personnel, and the choice between technical out-sourcing (technology transfer or subcontracting, for example) and the internal production of the different technical components of the expected product or process.

In technological cooperation, on the other hand, the more important issues are the quantity and quality of the resources (technical and nontechnical) that each partner will bring to the alliance and the ownership of the future results of the collective project. These issues are dealt with through complicated negotiations, and technological transactions between partners typically culminate in written agreements.

Table 1
Strategic European alliances by sector and rationale, 1980–89

Technology sector	Number of alliances	Risk, cost	Lack of resources	Technical compatibility	Accelerating innovation	Sharing R&D costs	Access to market	Gaining technical knowledge
				Main reasons for creating alliance (percentage of totals)				
Biotechnology	847	1	13	35	31	10	13	15
New materials	430	1	3	38	32	11	31	16
Information technology	1660	4	2	33	31	3	38	11
Automobile	205	4	2	27	22	2	52	4
Aviation, defense	228	36	1	34	26	nil	13	8
Chemicals	410	7	1	16	13	1	51	8
Consumer electronics	58	2	nil	19	19	nil	53	9
Food and beverages	42	1	nil	17	10	nil	43	7
Heavy electrical equipment	141	36	1	31	10	4	23	11
Health instruments and technology	95	nil	4	35	40	2	28	10
Others	66	35	nil	9	6	nil	23	8

Source: The Economist, 27 March 1993.
Note: Multiple choice question.

Memorandums of understanding (MOUS) are the most frequent way of institutionalizing the new routines within the cooperating firms and other members of the alliance. In a few cases, R&D joint ventures are created as semi-independent corporations.

Designing these MOUS is a difficult and painstaking task that can take months and even years. Since most of the resources and results are intangibles (skills, patented and non-patented knowledge of product and process technology), the precise evaluation of the research assets and results is particularly difficult; negotiations can collapse before the parties arrive at a written agreement. The same risks are present during collaboration, when unexpected results are often achieved. However, firms can *learn* to cooperate, in a process similar to productive learning (learning by doing), and the marginal transaction cost of each MOU and each alliance decreases over time.

Internal R&D is still necessary for firms involved in technological alliances. Since in the end all firms prefer to own some exclusive proprietary knowledge, cooperative projects conducted with external units are typically developed in parallel with purely in-house projects. New routines are added to, but do not replace, existing ones. The R&D organization becomes more complex, and the strategy used for protecting intellectual property becomes diversified, because some research results are jointly patented or protected by secrecy as the common property of all cooperating partners, while others are appropriated in the traditional ways used by the firms before the alliance.

Horizontal Versus Vertical Alliances

Technological cooperation between vertically linked firms has been observed in a nonsystematic way for many years. Electrical utilities and process industries, for instance, have collaborated for decades with their contractors and suppliers.[2] User-producer innovation was described more systematically in the management literature of the late 1970s and 1980s, mainly in the works of Von Hippel (1977) and Lundvall (1988).

Technological interaction and cooperation between rivals (firms operating in the same industry) was also occasionally observed by business historians (Hounshell and Smith 1988) but was not noticed, until recently, even by leading management theorists (Porter 1980, 1985). Both types of cooperation were noticed in the business administration

2 For a detailed account of the collaboration between Ontario Hydro, Atomic Energy of Canada, Canadian General Electric, and other firms in the design of the CANDU reactor, see Bothwell (1988).

literature in the late 1970s by Von Hippel (1977) and then, in the early 1980s, by Fusfeld and Haklisch (1985) and others. Technological cooperation among firms operating in the same industry was harder to admit, because of the stringent antitrust regulations prevailing in the United States. But in 1984, the American government modified its antitrust legislation in order to facilitate technological cooperation among rivals, and the new legislation attracted some attention from academia. In fact, as we shall see in chapter 2, technical cooperation was already widespread in the United States, as well as in Canada, either through industry associations or through industry-university research consortia (Fusfeld and Haklisch 1987).

National Versus International Alliances

National technical alliances are different from international alliances. National alliances consist of a larger proportion of universities and public laboratories; they usually have more members and a larger proportion of small- and medium-sized enterprises. Their goals are more likely to be either fundamental, or basic, research than development. Public funding and, sometimes, the public initiation of alliances are widespread. The budgets of these alliances are smaller than the budgets of international partnerships. Industry-university research consortia are typical of these alliances.[3]

International alliances, on the other hand, are typically initiated by private firms. The firms are usually large, or at least medium-sized, and there is usually little university or public laboratory involvement. Except in Western Europe, which has the European Union (EU) programs (see chapter 2), public funding is scanty. The development of new or improved products and processes for the market, not preliminary, basic, precompetitive research, is the usual goal of international alliances. Since product development is more costly than the precompetitive research, partnership budgets are, accordingly, much more important. Collaboration usually results in commercial products or processes, so agreements often include production, application, and/or commercial clauses about using the results of R&D in manufacturing and marketing. Most large multinational corporations, including Canadian-owned and controlled ones, are involved in international alliances (Perlmutter and Heenan 1986; Mowery and Rosenberg 1989; Niosi and Bergeron 1992; Gugler 1992).

3 The Center for Integrated Systems (CIS), based at Stanford University, is an example. Founded in 1980 and funded by some 20 industrial partners, CIS conducts research on semiconductor design and fabrication.

TOWARDS AN EVOLUTIONARY THEORY OF FLEXIBLE INNOVATION

Innovation is supposed to be a key component of the strategic behaviour of the firm. Through technical innovation, firms create new products and processes that give them temporary monopolies and the associated high profits. Since on this Schumpeterian approach, firms conduct in-house R&D programs to create exclusive competitive advantages, there is little room for interfirm cooperation. But the building blocks for a theory of cooperation can be found in some new ideas developed in the last ten or twenty years by heterodox economists and historians of technology. The following is a brief review of these building blocks.

Evolutionary Economics

Evolutionary economics provides a general framework for understanding technical alliances and cooperation. In the course of their life, firms acquire a specific set of operating practices or *routines* that change slowly under the influence of changes in the economic environment (Nelson and Winter 1982). It can be argued that technological alliances are themselves routines adopted by firms under internal and external constraints and changing under the influence of a dynamic environment.

Managerial evolutionism brings other fresh elements into the analysis as in biological evolution, organizational forms and routines like technical alliances are learned, developed by trial and error; *variation*, *selection*, and *competition* explain the patterns of change (Saviotti and Metcalfe 1992). Different learning processes create the variation, the new specialized capabilities, technologies, and routines that are then selected by the market or by government agencies.

Like transaction-cost analysis, evolutionary economics, is based on the assumption of the *bounded rationality* of economic agents, since this assumption is necessary for understanding learning processes of all sorts, including R&D processes such as technical alliances (Simon, Egidi, et al. 1992). Most importantly, the evolutionary perspective maintains that economics must be, like all contemporary natural sciences, an *historical* science and that trajectories – including organizational and technical trajectories – matter. In other words, the neoclassical paradigm of the nineteenth century, which was modelled on Newtonian physics, has to be replaced with the new scientific approach of the twentieth century, based not on an immutable universe but on a world in perpetual change, where equilibrium is only one dimension of dynamic systems. At the end of the twentieth century, big bangs, drifting continents, and biological mutations carry the day in the natural sciences; economics should adapt itself to contemporary science and adopt a more historical perspective.

The New Techno-Economic Paradigm

Technological cooperation is linked to the emergence of a major new phase in the development of technology. Freeman and Perez (1988) classify innovations into four categories based on their increasing complexity and systemic effects. The categories are: incremental innovations, radical innovations, changes in technology systems – the clusters of radical and incremental innovations involved in the introduction of synthetic materials are an example – and changes in techno-economic paradigms. In the last category, the changes involve many clusters of technology systems together with *organizational* and *managerial* innovations.

Changes in techno-economic paradigms can be related to each Kondratieff wave of prosperity. The introduction of steam power and the railway from the 1830s to the 1890s was one such change: it brought mass production and the first modern corporations, and it fostered the increased production of electricity, steel, coal, oil and gas, and synthetic dyestuffs. The increased production of electricity and the accompanying birth of the automobile, heavy engineering, and petrochemical industries had the same effect from the 1890s to the 1940s. The corporate form of organization was extended to most industrial, commercial, and financial sectors. Taylorism, a new style of management, emerged and was diffused through the new, large enterprises. National oligopolies and monopolies developed in North America and Western Europe.

The next techno-economic paradigm, developed since the 1940s, involved innovations in the production of automobiles, airplanes, consumer durables, and synthetic materials. The assembly line model of production became generalized; the multinational corporations, with their international hierarchies, dominated the economy.

A new techno-economic paradigm began to emerge in the postwar period, and it gained momentum in the 1980s. Freeman and Perez (1988) call it the "information-technology paradigm," since it is characterized by the general use of computers and numerically controlled machines and the application of optical fibres and the new ceramics to information services, robotics, and the production of new goods. New technologies, namely, biotechnology and advanced-materials technology, also appear. Horizontal communication, collaborative research among independent organizations, and concurrent engineering displace the previous vertical and sequential links between organizations and between functions within organizations. There is a massive entry of smaller, innovative firms and a rapid diffusion of technology that brings increased uncertainty and risk. Collaborative research appears as the effect of this increasingly turbulent environment, first in information technologies, then in other new technologies; the new paradigm finally catches up with established firms of the previous techno-economic paradigm.

This approach suggests that the new technology has resulted in widespread innovation by organizations seeking to integrate technological change into efficient production systems: innovations include horizontal communication in the firm, concurrent engineering, flexible production, and technical cooperation. In a context of rapid technological change, increased complexity, and the massive entry of competitors into the market, collaborative research may reduce the risks, uncertainties, and costs associated with R&D in information technology.

While Freeman and Perez have correctly put the accent on electronics, the new techno-economic paradigm must now include the two other generic, though less-developed, technologies just mentioned, advanced materials technology and biotechnology. The three of them constitute "information-intensive production systems" (Willinger and Zuscovich 1988): the technologies of production are more closely linked to scientific advance than ever before. This new technological paradigm is characterized by a qualitative leap in complexity requiring an increasing number of specialized, advanced knowledge-inputs from different sources.

Localized Learning: Technology as a Quasi-Public Good

Before we can develop a theory of flexible cooperation, we must offer a critique of the neoclassical concept of technology as a codified, accessible, and divisible public good. In neoclassical theory, technology is perfectly codified information, easily transmissible between independent firms, which, consequently, do not need cooperative R&D. This neoclassical perspective has been challenged from several angles. In several seminal articles, Stiglitz (1987) has insisted that technological learning is mainly localized. He argues that technology is different from science in that it is much more specific to particular industries, processes, and products. As a result, technology is much more likely to become obsolete than science, and much localized technological progress has no impact on (and no utility for) other industries and technologies.

It follows that R&D, which is, as we have seen, a specific type of learning process, is also mainly a localized activity. From Stiglitz' perspective, we can deduce that cooperative research is often so specific that the results may be applied only by the partners themselves, with few potential spillovers to imitators and competitors. When a large manufacturer of central telecommunications switching systems and a major cellular telephone producer conduct a collaborative research project to make their equipment totally compatible, the result of such a project is useful only for them (see chapter 3). Consequently, the protection and appropriation of this intellectual property should be easy for members of the alliance, since few externalities and spillovers can be obtained by nonmember firms.

Further, since technology mainly involves specific knowledge, it is also best characterized as only a "quasi-public" or an "impure" public good. The transmission and diffusion of technology depends both on the stock of knowledge of the potential buyer or imitator and on the degree to which the technology is explicit, codified, and incorporated into manuals, documents, drawings, and blueprints. But again, several authors insist that, in fact, technology mainly involves implicit, tacit knowledge that is incorporated in the expertise of workers, technicians, and engineers (Cohendet, Héraud, et al. 1992). The apparent paradox of technical alliances between complementary but competing firms thus disappears: cooperative knowledge production requires close collaboration between the bearers of the implicit, tacit knowledge: the R&D personnel of alliance members.

Several economic historians have drawn similar conclusions. Instead of understanding technology as public information, they consider that it "might be more usefully conceptualized as a quantum of knowledge retained by individual teams of specialized personnel" (Rosenberg and Frischtak 1985, vii). This knowledge is, in their view, mostly tacit and acquired through productive experience as much as through formal technical education.

From Flexible Production to Flexible Innovation

The new manufacturing is based on flexible production. A major contribution of the French regulation school is the light it has shed on the transition from Fordist mass production, which flourished from the 1920s to the 1970s, to flexible manufacturing since then. In a context of opening markets, new industrial competitors, and increased economic turbulence, rigid, large-scale, Fordist production is a hindrance, while flexible production produces strong advantages. The new electronic technology, including numerically controlled machines, robots, computer-integrated manufacturing, and local-area networks, has made flexible production feasible for the industrial firm.

The same reasoning applies to technological production, research, and development. In a turbulent technological environment with several new competing technologies emerging, technical alliances foster "flexible innovation." Collaborative research permits a more rapid switch from one technology to another and, compared to mergers and acquisitions, makes it easier to incorporate complementary knowledge, with reduced costs and risks.

Flexible organizations like alliances for R&D may be more effective at producing the intended results and more efficient, less wasteful of resources, than more traditional hierarchical, organizations. By combining different physical, financial, and human resources from a wide array of economic agents, alliances may procure an acceptable solution to the partners' technological problem, whereas each company alone may not be able to find any solution at all. In the complex environment of modern technology, this may be a key advantage of alliances over purely in-house, more traditional R&D. Furthermore, fewer resources are required for technological alliances than for mergers and acquisitions, since alliances do not force one company to absorb unwanted assets of another (Bleeke and Ernst 1993). And they permit firms to conduct new research projects without obliging them to acquire expensive laboratories or hire additional R&D personnel that may prove useless in subsequent projects. In other words, since alliances may reduce sunk costs, they may be an efficient way of organizing innovation.

Organizational Ecology

The ecological approach to organizational dynamics is another important ingredient in the new theory of flexible innovation. Founded by Michael T. Hannan and John Freeman (1977) and Howard E. Aldrich (1979), the organizational-ecology school has produced many theoretical and empirical studies that combine nicely with previous approaches. As Sidney Winter (1990) points out, organizational ecology is closely related to the evolutionary perspective, with which it shares the assumption of bounded rationality and a skepticism toward highly intentionalistic explanations of organizations. However, evolutionism has placed more emphasis on long-term changes within the organization, while organizational ecology has been more attentive to the links between the firm and its environment.

Organizational ecology brings concepts like the population of firms, their density, market entry, exit, and mortality. It also brings useful theories about age dependence (new firms tend to show higher mortality rates than older ones) and resource dependence (younger firms tend to starve for resources because of their tenuous links to their environment). The hypothesis of *protective alliances* between large and small firms, in which the former bring temporary resources to the latter, is particularly useful for analysing new industries and technologies, like biotechnology and advanced materials, where small innovative enterprises compete with large corporations that are buffered by their previous production from start-up costs, risk, and uncertainty.[4]

National Systems of Innovation

Technological change is increasingly seen to result from the interaction of a vast array of innovating units (private and public firms, government research laboratories, universities) supported by government financial agencies, venture capital firms, educational and regulatory bodies, and the like (Freeman 1987, 1988; Lundvall 1988, 1992; Nelson 1988; Niosi, Bellon, et al. 1993; Niosi and Bellon 1994). These different units are linked through a complex web of informational, personal, financial, commercial, and other ties. Together they constitute the "national system of innovation." The interaction of the innovating and supporting units produces the variety of learning processes that is the precondition for successful technological change.

From this perspective, the focus is no longer on individual firms but on the systemic effects of interaction. The theory is that some national systems of innovation (NSIS) are more effective than others in producing technical change, either through better networking or through a better mix of the different kinds of innovating and supporting units. Alliances are only one form of networking and so represent only part, but a central one, of the national system of innovation. With both alliances and NSIS, the focus in the economics, politics, and management of technology shifts from the firm itself to the whole array of relationships that provide the innovating firm with regulation, information, skilled labour, R&D funds, and the like. Technology management is no longer exclusively centred on what the firm does within its own boundaries but also on how it manages its relationship with its environment.

4 See the excellent organizational-ecology analysis of the semiconductor industry by J. Freeman (1990).

Building New Concepts

To formulate a new theory of alliances, we must not only rearrange and develop some of the most useful existing theories and concepts, we must also create new ones. Some of the most important new ones are the concepts of technological cooperation (versus competition), collective learning (versus learning-by-doing), and traded externalities.

Bringing Cooperation Back. Economic and management theories have devoted many pages to the advantages of competition and the costs of monopoly (Yoshida 1992). But the disadvantages of competition have seldom been analysed, especially as compared with cooperation. The various forms of economic cooperation have not been studied, and the optimal mix, in different markets, between competition and cooperation (on products, technologies, organization, and the like) has been neglected. Because of the increasing importance of cooperation, the comparative study of the costs and benefits of competition and collaboration may bring a major renewal to economic theory. Since economic efficiencies may result from cooperation, as well as from competition, a comparative analysis of both types of arm's-length transactions promises worthwhile results.

Collective Learning. In 1962, Kenneth Arrow published his classic discussion of learning-by-doing, a concept that highlights the productivity gained within the firm when the work force acquires a better grasp of newly incorporated technology. A few authors (Dutton and Thomas 1985; Siggel 1987; Teubal 1987) have elaborated Arrow's basic concept in an effort to understand other types of learning in the firm: learning by studying (in-house R&D), learning by consulting, and learning to invest. But it is now necessary to develop a new concept, which I shall call "collective learning," if we are to understand the type of cooperative, interactive learning that involves several economic units and collective projects. This new concept will be based on the study of formal and informal cooperation in R&D in four industries.

"Collective learning" is in some ways similar to "collective invention," a concept proposed by Allen (1983) to analyse technological developments in the British and American iron industries during the nineteenth century. Collective inventions and innovations are often the outcome of formal or informal cooperation, just as collective learning often takes place through the interaction of independent firms and other research units. There are, however, some major differences between the two concepts. Collective invention is based on a free exchange of information. Collective learning is a more systematic process, carefully planned and monitored by members of technological partnerships. Collective invention makes information available to all firms in the industry. But in collective learning, only the members of the alliance will capture the increase in capabilities produced by the partnership.

Trading Externalities. The theory of externalities is also an underdeveloped port of traditional economic theory. A technological externality exists when some activity of party A imposes a cost or a benefit on party B for which A is not charged or compensated by the price system of the market economy (Whitcomb 1972). In economic theory, externalities, be they positive or negative, are *imposed* by one economic agent on another. Externalities are spillovers, social returns, unintended benefits that the producer cannot capture for itself (Mansfield 1977). Or they are costs (such as environmental disruption or pollution) that one company imposes on another. But an alliance involves a deliberate, conscious search for mutual externalities by all cooperative partners; alliances involve a purposive exchange of positive externalities between firms and other parties. Alliances create a kind of trade in technological externalities: "traded externalities" are thus different from traditional "positive externalities," the kind of spillover that the economic theory of technical change has dealt with time and again in empirical analysis.

Learning processes, the accumulation of intangibles, and the control of externalities are well-known activities in high technology. In high technology, the firm's most important asset is knowledge, and most companies have some experience in fostering, auditing, and controlling the flow of knowledge across their boundaries. Confidentiality agreements between firms and confidentiality clauses in the contracts of R&D scientists, are two of the methods commonly used to deal with these potential, unwanted leakages. It is mainly in knowledge-intensive industries that companies are able to share some externalities through cooperation, while suffering (or benefiting from) the more traditional spillovers of information.

CONCLUSION

No established theory, either in management theory (at the firm level) or in industrial and evolutionary economics (at the industry level), can explain the present wave of technological alliances. However, it is possible to build a new theory by drawing a set of concepts and explanations from the present literature and by introducing or forging new concepts. From industrial economics, the most important concept for our purposes is the idea of R&D economies of scale and scope. From management science, the most important explanations are based on the firm's need to capture and accumulate intangible assets from users and other firms with complementary knowledge, to introduce new management schemes, and to reduce the risk and uncertainty associated with accelerated technical change and a more open competitive environment.

Transaction-cost theory has tended to predict either that interfirm arrangements are marginal or that they are transitional forms tending towards more classical, hierarchical forms of organizing the R&D function. But in this chapter, I have sketched a theory of flexible innovation that, for the first time, can explain why alliances are central to the strategy of the firm. Drawing on evolutionary economics, this theory understands technical change as a dynamic process involving firms strongly embedded in an environment that provides them with fresh knowledge, new routines, and other key elements required for survival, adaptation, and change. The environment (both economic and political) selects the organizational forms and the associated technologies that will survive. The frontiers of the firm thus appear to be relatively porous, since technical and organizational information, not simply knowledge of price and output, flows between the enterprise and its environment. On this theory, technological alliances appear to be a specific form of technical cooperation, developing simultaneously with information-intensive production systems: in electronics, biotechnology, and advanced materials. Firms cooperate within a turbulent technical environment when they need closely coordinated, flexible technological innovation.

Technological collaboration forces us to reconsider the nature of technology, which now appears to be much more specific, tacit, and more deeply embedded in the expertise of individual agents than is usually admitted. Consequently, learning processes appear to be much more localized, and technological development appears to need closer collaboration than was previously thought. Thus it becomes clear that alliances are interactive, evolutionary learning processes.

Technological alliances are situated in a middle ground between informal interfirm cooperation and outright merger and consolidation of technological assets (Table 2). Technological cooperation, including the formation of alliances, is a key part of a wider set of linkages and relationships (informational, personal, financial, regulatory, and the like) between innovating and supporting units. Together, these relationships constitute a national system of innovation.

The rapid and widespread development of technical alliances in manufacturing raises new questions about the sources of innovation,

Table 2
Technical cooperation

General description	Informal interfirm cooperation	Strategic technological alliances	Mergers
Specific legal forms	Exchange of information Industrial technical conferences and journals	MOUS Industry/university consortia R&D joint ventures	Total or partial coordination of all R&D activities

the appropriability and protection of the R&D results, and the organization of the R&D activities within the firm and between the enterprise and external units, questions to which no present theory can completely respond. Our goal in this empirical research is to clarify these questions and, I hope, provide new answers.

2 A Cross-Country Comparison of Government Policy

Governments have financed technological innovation since the second half of the nineteenth century, when the links between science, technology, and economic development became evident and when the risk and uncertainty of investment in technical change was understood.

There are many specific arguments for government support of investments in R&D. First, investments in the production of knowledge produce uncertain returns that are not entirely appropriable (Arrow 1962). Since knowledge is, at least partially, a public good, a firm can take exclusive possession of the benefits of an innovation only when it has a perfect monopoly. Basic knowledge is particularly difficult to appropriate, and its practical applications are not always immediately evident to private enterprise. As a result, most fundamental research is conducted with public support in universities and government laboratories.

Further, since some R&D is too expensive or too risky to be conducted by individual firms, governments share the costs and risks of large-scale projects: projects in the aircraft and aerospace industries are examples.

Governments fund both the technological development of infant industries, because entry would be difficult in the face of international competition, and the technical reconversion of declining industries, in order to suppress disruption and collective losses. In some industries, such as agriculture, where production is typically conducted by small firms that cannot allocate sufficient funds to R&D, governments tend to conduct in-house research, agricultural research, for example, which is then disseminated to private farmers (Coombs, Saviotti, et al. 1987). Governments also support technical change in areas of collective benefit, such as defense, telecommunications, the environment, public health, and alternative energy, because no enterprise would easily benefit from private R&D investments.

The most important argument for government support is that basic scientific knowledge is a quasi-public good: its use by one economic agent does not preclude its use by another, since knowledge can be shared with small cost and no loss. Thus there are important economies in the public funding of basic research (Dasgupta and Stoneman 1987).

Governments try to promote national competitiveness which, in modern economies, is increasingly driven by knowledge-intensive manufactured products. While the trade in these products has increased rapidly in the last few decades, resource extraction and transformation are now linked to underdevelopment, since the relative prices of raw and semiprocessed materials have declined and the competition from less developed countries is strong in these products whose added value is low. The new knowledge-intensive products require a public scientific and technical infrastructure and increasing technical change in the private sector.

We can see, then, that governments fund technical change to cope with market failure: in many cases private enterprise simply cannot allocate enough resources to investments in R&D. But governments may also support the diffusion of innovation. Since technical knowledge is a public good, it may be underutilized if it is appropriated by a single firm. So government finances cooperative research in order to promote the diffusion of results, as well as to foster complementarities among corporations and avoid the duplication of financial support.

PUBLIC FUNDING OF COOPERATIVE RESEARCH

Since the beginning of the twentieth century, governments in several industrial countries have financed some cooperative research. In the United Kingdom, for instance, trade associations first created research laboratories in 1918, and because of the small size of the enterprises involved, government usually financed half the costs of the cooperative research, which was more common in traditional industries (Norris and Vaizey 1973). More generally, trade associations have sponsored collective fundamental research of common interest and have received some public help in the United States, Western Europe, and Canada (Fusfeld and Haklisch 1987).

A few large-scale government projects, in which huge costs and high risks discourage private enterprise, have also received public subsidies for concerted action. The Manhattan project in the 1940s, which the United States considered crucial for its national defense, and the design and construction of the CANDU reactor in Canada in the 1950s are examples of government support for technical collaboration between private and public enterprises.

Now, however, governments of all the industrialized countries are fostering technological alliances in high-technology industries for projects of all sizes in all industries. The explicit public goals vary, but they usually include the following:

1 *Diffusing technology.* Governments have recognized the need to improve the diffusion of new technology through the cooperative acquisition of technology and through cooperative research. The Japanese experience with collective technology imports and production under the guidance of the Ministry for International Trade and Industry (MITI) is often cited as an example of the benefits of government intervention in technology markets.

2 *Avoiding duplication.* Rival research units may select similar projects and multiply applicants for research funding (Dasgupta and Stoneman 1987); it may then be efficient for governments to reduce overall R&D costs by avoiding the duplication of subsidies. Governments prefer to fund a small number of consortia instead of many individual enterprises.

3 *Accelerating technical change.* The triple generic technological revolution (electronics, advanced materials, and biotechnology) made it necessary to combine complementary, complex skills. Governments are promoting the development of cooperation in order to generate R&D spillovers into traditional industries when new technologies are adopted.

4 *Covering transactions costs.* Alliances, and technical cooperation in general, involve larger transaction costs than purely in-house R&D. Governments may support collective innovation by assuming some of these collective transaction costs (Watkins 1991).

In other words, governments promote technological alliances in high-technology industries for the same reason they support technical change in individual industries: market failure. Like other economic agents, private firms suffer from bounded rationality. Private enterprise does not always understand the advantages of technical cooperation or the nature of the present technological revolution. Private enterprise avoids risks (including the free-rider risk implicit in technological cooperation with external partners) and costs (including the transaction costs implicit in alliances). But isolated firms with older methods of production may well be unable to cope with international competition or enter into the new industries generated by new technologies.

Countries have promoted cooperative R&D in various ways. In the immediate postwar period, Japan probably set the pace, followed by the European Economic Community, the United States, and then Canada. As we have seen, there are traces of cooperative R&D by private firms and industry associations in many different developed countries well before 1945. But what is new in the postwar period is the acceleration of the trend in the private sector and the transformation of technical cooperation into the major goal of government innovation-policy in the 1980s.

Japan

Cooperative R&D in Japan is probably as old as Japanese industrial policies that sought to catch up with the West. Since the Meiji revolution of 1867, the Japanese government has pushed domestic companies to look for technical assets in the West and to cooperate to maximize the spin-offs from technology transfer from abroad. Hard necessity drove the Japanese government to foster cooperation in technology transfer in order to reduce the currency costs of transfers and insure a better diffusion of imported technology (Oshima 1984). In the late nineteenth and early twentieth century, the formation of Japanese conglomerates (the *zaibatsu*) helped the development of technical cooperation among corporations within the same group (Blackford 1988), and during the 1930s, the cartelization movement, backed by the Japanese government, also helped technical cooperation among competing firms.

Antitrust considerations, such as worries about the risk of abusive collusion or merger among technical cooperators, were never a major deterrent to technical collaboration in Japan, or in Europe for that matter. The predominant goal of government policy was always to attain a critical research mass and catch up with the West. In the postwar period, technical cooperation at the industry level started in the electrical utilities and soon spread, under the aegis of MITI (Johnson 1980; Okimoto 1989), to capital-intensive and then high-technology industries. The promotion of technical collaboration by private firms on their own and by private firms with government laboratories and universities was just one of the many measures implemented by the Japanese government. One major landmark was the Research Associations for the Promotion of Technology Act of 1961, which established the engineering research associations. Dozens of research consortia, supporting projects on everything from chemicals to aircraft to ceramics, were created under this legal umbrella. MITI is not the only government agency involved in these cooperative activities. The Nippon Telegraph and Telephone Company, the Ministry of Education, Science, and Culture, and the Ministry of Posts and Telecommunications often take the initiative and provide financial support for technical alliances (Levy and Samuels 1991).

Microelectronics provides a striking example of these policies (Majundar 1988; Okimoto, Susano, et al. 1984). In 1957, in order to develop a modern microelectronics industry, the Japanese government passed the Electronics Industry Development Emergency Law, which created an electronics council within MITI to develop an annual electronics development plan. The resulting annual plans included technical targets, government allocations for cooperative research (created "to minimize duplication of efforts and conserve R&D resources"), and exemptions from antitrust laws for collaborative research. From 1958 to 1970, MITI's direct financial support for a microelectronics industry amounted to 4.3 billion yen, the equivalent of 45 percent of all direct MITI subsidies for technical innovation.

As this example and several others show, the Japanese approach is one in which cooperative funding of R&D is only one part of a set of policy measures. Okimoto summarizes them as follows:

1 Central coordination and consensus building, including agreements on targets and ways of achieving them, the division of labour between firms, government and private laboratories, and information and communication flows between all partners;
2 Supply-side incentives, including tax credits for R&D and preferred loans, subsidies, and the like to consortia;
3 Demand-side guarantees such as public procurements to members of publicly backed collaborations;
4 Infant-industry protection through import duties, control over foreign direct investment, and control over the purchase and licensing of foreign technology; and
5 Legislation providing selective exemptions from antitrust laws for partners backed by government programs. (Okimoto, Susano, et al. 1984, 99–100)

The European Union

Governments in Europe began to promote industrial and technical cooperation in the late 1950s and early 1960s in an attempt to catch up with the United States in high-technology industries. Government corporations and research laboratories were the main actors in these early initiatives. CERN (Centre européen de recherche nucléaire) was created in 1954, and other initiatives followed: the European Launcher Development Organization (ELDO) Agreement (1961), forefather of the European Space Agency (1975), and Arianespace, the British-French agreement on Concorde (1962), forefather of today's Aérospatiale. Also, private intercompany agreements in high-technology industries, the computer industry, for example, flourished between 1961 and 1973, with government legal help and financial support (Jéquier 1974).

A massive trend towards the public promotion of technical cooperation in high technology began in Europe in the early 1980s, when the EC launched a handful of technology programs promoting cooperative R&D among private firms, government laboratories, and universities of different EC countries. The main programs included ESPRIT, BRITE, RACE, EURAM, JESSI, Biotechnologies, and JOULE.[1] To qualify for support, a project had to bring together partners from at least two different countries in the EC. The programs put billions of public dollars into collaborative R&D and soon produced thousands of inter-European research projects. The original, more basic, precompetitive focus of collaborative research changed through the years towards more applied development projects (Mytelka 1990, 1992).

The first phase of ESPRIT, for example, was launched in 1983 and completed in 1987. It had a total budget of Can$1.1 billion, on the basis of which it financed some 225 projects at a 50 percent rate. ESPRIT I was centered on precompetitive research, but its second phase, ESPRIT II (1988–92), has a budget of Can$2.3 billion to fund several hundred R&D projects with a higher emphasis on applied research and development. The RACE program (1987–92) has a budget of Can$700 million to fund 50 percent of several hundred projects.

Another major intergovernmental scheme, launched in 1985, is EUREKA, in which nineteen European governments are involved, and

---

1 The full names of the programs are as follows: European Strategic Program for Research and Development in Information Technology (ESPRIT); Basic Research in Industrial Technology for Europe (BRITE); Research and Development in Advanced Communications for Europe (RACE); European Research in Advanced Materials (EURAM); Joint European Submicron Silicon Integration (JESSI); and Joint Opportunities for Unconventional and Long-Term Energy Supply (JOULE).

where foreign partners (including Canadian partners) are admitted for specific projects (Braillard and Demant 1991). EUREKA first associated Austria, Finland, Norway, Sweden, and Switzerland with the twelve EC countries, and then added Turkey and Iceland. By mid-1990, EUREKA represented more than 2000 corporations, government laboratories, and universities on 385 R&D projects with a total budget of 7.8 billion ECUs. The majority of the projects were in the fields of the environment, robotics, biotechnology, and information technologies.

The European commitment to cooperative research seems much more comprehensive than the Japanese, both in funding and in scope (number of industries and types of partners involved). The final goal is to create pan-European networks of innovators that will foster competitiveness, mainly in areas where European corporations are lagging (Claverie 1991), and help Europe catch up with the United States and Japan in critical technologies.

In parallel with the EC programs, most Western European governments have developed national programs for cooperative research within their borders. The Federal Republic of Germany launched a five-year, Can$1.2 billion program in 1984 to reinforce domestic capabilities in microelectronics, computers, and telecommunication equipment. This scheme followed the famous five-year Alvey program, launched in 1983 to promote the British microelectronics industry with $500 million in government funds. Most European governments followed suit, fostering several high-technology industries while they continued or increased their participation in the EC programs.

A thorough and very positive assessment of the impact of the EC programs on the R&D system in France was published in 1990. The report showed that the programs had accelerated collaboration among French and other European partners and had accelerated innovation and technical training and fostered scientific publication, while leaving to the partners the choice of nearly 90 percent of the project subjects. Nevertheless, the collaborative R&D produced a surprisingly small number of patents (Larédo and Callon 1990).

The United States

In the United States, as in Great Britain, cooperative research was at first a private endeavour linked to trade associations and traditional industries (Fusfeld and Haklisch 1987); the research was mostly precompetitive. The Institute of Paper Chemistry (created in 1929) and the Textile Research Institute (1930) provide good examples of earlier collaborative R&D. Much of it was located in universities and funded by those universities, industry, and government. The latest of these traditional institutes are the Electric Power Research Institute (EPRI, 1972) and the Gas Research Institute (GRI, 1976).

Present-day cooperative research in the United States focuses on the development of high technologies; the goal is mainly applied research and development. Government programs increasingly support cooperative research. This new trend started in 1980 with the development of the Stanford University Center for Integrated Systems (CIS). Among the most important that followed were the Semiconductor Research Corporation in 1982 (SRC), the Microelectronics and Computer Technology Corporation (MCC, 1983), Bell Communications Research (Bellcore, 1984), and the Semiconductor Manufacturing Technologies Corporation (Sematech, 1987). By 1987, there were approximately 70 of these consortia in American industry, some, like Sematech, cofinanced by the U.S. government, others, like MCC, relying basically on private contributions of members. Some, like SRC, were focusing on basic research, while others, like MCC, concentrated on applied research and development.

Parallel to the development of large public consortia are the thousands of new technical alliances developing through private agreements and joint ventures between private firms of all sizes, generally with little publicity and without public support. The announcement in 1991 of a major technical alliance between IBM and Apple is an example of the new type of cooperative research that is spreading in the United States.

However, the American government is not initiating these organizational innovations in R&D in the way the Japanese or the European governments are. One main contribution of the American government has been to reduce the legal barriers to cooperative research through a 1984 amendment to the antitrust legislation (White 1985). Some consortia are financed by the Advanced Research Projects Agency (ARPA), the Department of Defense arm created in 1958 for funding high-technology R&D initiatives. From the $1 billion annual budget of ARPA, some funds have found their way to advanced technology consortia like Sematech. Other collective research projects are funded directly by the U.S. Department of Defense (DOD) with a nearly $40 billion annual budget for research, development, testing, and evaluation; however, the goals of the DOD projects are purely military: the development of missiles, advanced aircraft, submarines, and the like.

A debate is currently raging in the United States over the relevance of military-oriented research, which has been geared towards low-volume, high-performance products regardless of cost, to the development of civilian products, where the loss of American competitiveness is most evident. Also, the usefulness of using military agencies to support commercial R&D is being debated. Furthermore, the public funding of the new research consortia is questioned in a White House document on U.S. technology policy sent to Congress in September 1990 (Executive Office of the President 1990).

We may be witnessing the emergence of an American technology policy different from the mission-oriented one that took shape in the postwar period. This new policy would be directed more towards commercial than military technology; it would be more supportive of cooperative than of isolated innovation and more centered on diffusion and downstream innovation than on basic scientific research (Branscomb 1991, 1992).

Canada

Technical cooperation with some government support is not new in Canada, but the intensity of the effort is. While the National Research Council (NRC) failed in its attempt in 1918 to form Canadian research guilds based on the British experience, several industrial associations did so on their own initiative. Cooperative research started at the beginning of the twentieth century on the initiative of either university or industrial associations. Hull and Enros (1988) give an account of some of the most remarkable cases since 1901. Collaboration between universities and industries started even earlier, during the second half of the nineteenth century, through research chairs, research contracts, and research consulting by university professors. The Pulp and Paper Research Institute of Canada (PAPRICAN), still one of the major performers in Canadian R&D, was founded in 1925 at McGill University with large industrial and some public funds.

The Canadian nuclear effort, which gave birth to the CANDU reactor, started right after World War II. It was a collective undertaking of the National Research Council and, since 1952, Atomic Energy of Canada, Ontario Hydro, and several private manufacturing and engineering firms (Doern and Morrison 1980; Lermer 1987). Since the 1940s there has also been some technical collaboration in the aerospace industry between the aircraft builders, Canadair and DeHavilland, and their major suppliers, particularly the engine manufacturers. The industry received government support for both in-house and collaborative R&D (Canadair 1985; Sullivan and Milberry 1989). In fact, the entire Canadian aerospace program, from the early 1960s to the present, has been conducted with collaboration between several national and foreign enterprises and laboratories, both public and private.

Public laboratories also fostered some networking between R&D agents: the federal government created the Fisheries Research Board in 1898 to help private industry. Also in the nineteenth century, the federal and provincial mining and forestry departments conducted technical research for industry. Since the foundation of the NRC in 1916, eight provinces have established their own research facilities: Alberta in 1921, Ontario and Manitoba in 1928, British Columbia in 1944, Nova Scotia in 1946, Saskatchewan in 1947, New Brunswick in 1962, and Quebec in 1969. Public laboratories collaborated with industry either by making their own research results available or by doing contract research and consultancy for industry (Eggleston 1978; Le Roy and Dufour 1983). This kind of "soft," informal cooperation is different from technical alliance, but it has paved the way for them.

Until the 1960s, and in parallel with these and other scattered activities, governments subsidized innovation within individual firms and universities, but not within consortia, alliances, or networks of firms (Tarasofsky 1984). Then, during the 1960s, the emphasis of government programs slowly changed to explicitly include financing for collaborative research and for consortia and industrial associations. The Industrial Research Assistance Program (IRAP), established by the NRC in 1962, and the Program for the Advancement of Industrial Technology (PAIT), established in 1965, were two of the schemes for supporting collaborative research.

In the 1980s, cooperation became the main goal of more than one hundred new programs. The Programme d'actions structurantes du Gouvernement du Québec, launched in 1984–5, was designed to create forty R&D networks of corporations, universities, and government laboratories, under the aegis of the Quebec Department of Higher Education, Science, and Technology (MESST). Ontario followed suit in 1986 with its own Centers of Excellence Program, which created eight networks with basically the same goal as Quebec's initiative. Then, in 1989, the federal government created, on the same principles, the $240 million, four-year, InnovAction Networks of Centers of Excellence Program, jointly managed by the Natural Sciences and Engineering Research Council (NSERC) and the Social Sciences and Humanities Research Council (SSHRC). InnovAction funded some sixteen networks of R&D units across Canada. All three programs funded university research projects with support from the private sector.

In 1984, partly for financial reasons, the federal government began promoting links between federal laboratories and industry, and in 1987 it ordered all federal laboratories to create advisory boards with industrial members. As a result, the NRC and other government laboratories began to emphasize industrial partnerships and collaborations

(NRC 1985, 1990). In the same year the federal government created the Industry, Science, and Technology Department (ISTC) by merging the Departments of Science and Technology and Regional Industrial Expansion. The new department is responsible for the Defense Industry Productivity Program (introduced in 1959) and the newly reimplemented Defense Industrial Research Program (1988). The Strategic Technologies Program (STP), another initiative of ISTC, has chosen three areas for special development: information technologies, biotechnology, and advanced industrial materials. The emphasis of these and other programs is increasingly on the funding of R&D alliances.

In 1988, the federal Department of Communications launched a five-year, $125-million information and telecommunications technology-development program. Vision 2000, one of the major Canadian R&D consortia, was formed under the aegis of this program (more on this in chapter 3). Then, international alliance programs were added. In 1990, the federal government, through Investment Canada and the Departments of External Affairs and Industry, Science, and Technology Canada started promoting links between Canadian companies and their European counterparts. The European Strategic Alliances Program has helped form those partnerships in the areas of computer software, electronics, biomedicine, ocean technology, opto-electronics, and industrial wastewater. Thirteen alliances based on that program were concluded in 1991 between Canadian companies and European counterparts; most were R&D partnerships.

The provinces followed or perhaps even anticipated this new cooperative trend. In 1986, Ontario launched the Ontario Technology Fund (OTF), a five-year $1 billion collaborative R&D program. It was followed in 1989 by the $300-million, five-year Quebec Technology Development Fund (Fonds de développement technologique du Québec). Both programs were intended to foster networks of innovative corporations and universities within the provinces.

It would be difficult to make an assessment, or even a complete list of the dozens of provincial and federal programs developed in the last five years to help create technical alliances among innovative organizations in Canada. By 1991 there were 129 federal or provincial programs supporting technical cooperation among firms, universities, and government laboratories. Quebec was the most active province, followed by Ontario and Alberta (Niosi and Landry 1993).

CONCLUSION

In the postwar period, the industrialized countries of Europe and Japan slowly developed policies supporting technical cooperation as a means to catch up with the United States. The trend took off, however, when, under pressures from globalized markets, accelerated technical change, and increased uncertainty, corporations active in high-technology research began to cooperate in R&D, and governments followed suit. Large government programs were created in both Japan and the EC to support collective innovation. Only later, did the United States and Canada adopt the new way of promoting R&D. In fact, the United States still lags behind Canada because of the established American tradition of leaving the market to decide about commercial technology – and the fundamental science and R&D in defense, infrastructure, and agriculture that supports it.

Although the benefits of cooperative research are generally admitted, concerns about government-promoted alliances have been raised, particularly in Western Europe. First of all, there are antitrust concerns that research partners may eventually merge and consequently increase economic concentration. Second, the programs have opportunity costs: public support for R&D on new technologies may deflect funds from other, less fashionable, but nevertheless important, research areas, such as public health. Third, in selecting collective projects, public agencies take the same risks that they take in trying to pick individual winners: the chosen project may well fail. Fourth, the transaction costs and risks of most EC and EUREKA programs of collaboration may be unacceptably high, because they are international projects involving several languages and legal frameworks and diverse commercial practices. Fifth, governments may encourage the creation of partnerships that are too large; in the case of ESPRIT I, it was shown that the average project had 5.5 participants, while the optimum seemed to be just 2. European experience shows that the greater the number of participants is an alliance, the greater are the problems of management and cost (Woods 1987). Finally, some critics have questioned national technology policy, particularly in the EC context, since government supported international alliances permit and even promote the flow of critical technologies across national borders (Soete 1991). Since the 1980s, in other words, by subsidizing international partnerships, national governments have been helping foreign rivals to improve their technical capabilities.

Others are sceptical that the model of research consortia typical of the Japanese system of innovation can be successfully generalized to apply in countries like those of Western Europe and North America (Levy and Samuels 1991). The Japanese economy consists of very stable institutional actors: Japanese firms do not merge, collapse, or disband as easily as they do in the West, and foreign direct investment is discouraged. This stability favours collaboration among Japanese firms instead of mergers or cut-throat competition. Also, Japanese firms accept government guidance and arbitration, and they trust government officials and public initiatives. Finally, the Japanese strategy of technological collaboration was progressively developed as part of a sustained effort to catching up with the West. All these elements are lacking in Western Europe and North America, particularly in the United States.

3 Electronics

WITH THE COLLABORATION OF
MARYSE BERGERON

The term "electronics" is at the same time the name of a new industry and of a new generic technology. The electronics industry emerged from the electrical industry as a result of a major series of innovations, popularly labelled the "microelectronics revolution." Its origins can be traced to the early postwar period when, in 1947, Bell Laboratories in the United States invented the transistor. Twelve years later, in 1959, Texas Instruments and Fairchild developed the integrated circuit. Then in 1960, Bell Laboratories introduced the electronic telephone exchange, thus launching the convergence of microelectronics and telecommunications. In 1969, the microprocessor was developed by Intel, and in a few years semiconductors took over the computer industry, then machinery manufacturing, and finally spread to most other industries, including those producing electrical and transportation equipment, scientific products, and the like (Forester 1988; Okimoto, Susano, et al. 1984; Queisser 1988). Today, the development of software for this increasingly sophisticated equipment has become a major part of what is increasingly called the information-technology industry.

ALLIANCES IN THE ELECTRONICS INDUSTRIES

Electronics is the industry where technical alliances have developed by far the most conspicuously in the last fifteen years. Starting with technology exchange and cross-licensing in the late 1970s, cooperation among producers became more widespread, complex, and significant during the 1980s. By the early 1990s, all large and medium-sized firms were conducting strategic technical alliances with other firms, either in the same country or across national borders (Fusfeld 1987; Langlois, Pugel, et al. 1988; Pisano and Teece 1989). All products of the electronics industry are covered by these alliances: semiconductors, telecommunications equipment, computers of all types and sizes, robots, printers, and other specialized instruments.

In the United States, large domestic consortia in the electronics industry include the Semiconductor Research Corporation (SRC), the Microelectronics and Computers Technology Corporation (MCC), the Microelectronics Center of North Carolina (MCNC) and the Stanford Center for Integrated Systems (CIS), all organized in the 1980s with industry funds and, in some cases, with government support. Some privately-organized domestic alliances are organized entirely within the electronics industry, for example, the Apple-IBM partnership, organized in 1991 to develop a new generation of personal computers. Others cut across industries, for example, the alliance in 1984 between General Motors and six small robotics corporations in the United States, organized in order to improve flexible operations in automobile manufacturing.

International alliances also prospered between European and American firms (the alliance of AT&T and Olivetti for the production of personal computers is an example from the 1980s) and between American and Japanese firms who wished to improve their semiconductors and process technology. Massive international technical collaboration in electronics appeared in Europe under the aegis of the ESPRIT, BRITE, RACE, and EUREKA programs (Delapierre 1991). More recently, strategic technological alliances have been organized between American, European, and Japanese partners. In July 1992, for instance, IBM, Siemens, and Toshiba announced an agreement to develop advanced memory chips for the computer industry of the twenty-first century: this billion-dollar R&D agreement is the largest technical alliance in the semiconductor industry to date. It covers only joint R&D costs, but it is expected to be extended eventually to joint manufacturing.

There were many reasons for this important reorganization of technological innovation in the world electronics industry. First and foremost was the desire to reduce the technological gap between the previous leader, the United States, on one side, and Japan and the EC countries on the other. American firms are still the leaders in the production of custom-made semiconductors as well as in microelectronics design and software and the manufacturing of supercomputers. The

Japanese lead in the production of memory chips and electronics process technology and in the application of microelectronics to robotics and numerically controlled machines (NCMs). While the EC countries are generally lagging in this industry, they compete successfully in several areas of telecommunications, robotics, and NCMs. This "levelling-off" of world technological capabilities, and the fact that different countries lead in different areas, has favoured international alliances.

Second, the acceleration of innovation in electronics has been dramatic. New products are regularly launched every year, rendering many others obsolete. The speed and turbulence of the process of innovation has forced firms to collaborate in order to reduce the escalating costs of R&D and acquire the requisite complementary knowledge in time. Finally, the development of common standards and the fact that a firm's products must be compatible with rival products has also forced companies to collaborate in R&D.

The present wave of technological collaboration probably appeared first and attained the highest level of development in electronics. As we shall now see, Canadian firms followed this trend and adopted the new pattern of innovation.

The Canadian Industry

The rise of the Canadian electronics industry after the war involved the transformation of a number of electrical office- and telecommunications-equipment producers, mainly foreign-controlled firms, IBM Canada, Northern Electric, and National Cash Register, for example, into electronics manufacturers. Several hundred Canadian-owned and controlled firms were incorporated as well, most of them producing semiconductors, specialized telecommunications equipment, and other electronic equipment for health and scientific purposes. And some large and medium-sized foreign-owned subsidiaries were absorbed by Canadian interests: the most outstanding example is, of course, the takeover of Northern Electric by Bell Canada in 1956.

Some sectors, notably telecommunications equipment and semiconductors for the telecommunications equipment industry, are conspicuously overrepresented in the Canadian electronics manufacturing industry. But producers of computers, computer parts, and other electronic equipment are underrepresented, except in specific niches; others, like producers of consumer electronics, are hardly represented at all. Along with their vertically integrated semiconductor producers, a few large and medium-sized telecommunications manufacturers dominate the Canadian electronics industry. In 1993, the telecommunications equipment industry had sales valued at $4 billion, while the

Canadian microelectronics industry, with only three manufacturers and 30 niche producers and designers, had an output of $600 million. A few large corporations are also active in flight simulators, computers, and other office equipment. All in all, in 1986 there were some 600 enterprises in the industry in Canada (SIC codes 335, 3361, and 3368), owning 683 manufacturing plants (Statistics Canada 1989). This was the target population for our research.

THE RESEARCH

Our research was originally intended to centre exclusively on the telecommunications-technology industry. However, the high level of interpenetration by, and technical alliances with other electronics sectors suggested that the whole electronics industry would be a better target for our research. Some of the largest semiconductor plants in Canada, for instance, are owned by the three largest telecommunications equipment producers. Moreover, the computer, satellite, and telecommunications-equipment industries are increasingly liked by product convergences, plant ownership, and technical alliances.

Characteristics of the Firms

The firms in our target population in this case, as in the following chapters, were chosen in accordance with the following criteria: they were (1) firms with Canadian operations, (2) firms involved in regular R&D, and (3) firms involved in technical collaboration either with Canadian or foreign partners or both. Both Canadian-owned and controlled firms and foreign subsidiaries conducting R&D in Canada were included. According to Statistics Canada (1991), there were 238 corporations performing R&D in the Canadian electronics industry in 1989. R&D expenditures varied from 3.6 percent of sales in business machines to 14.8 percent in telecommunications equipment.

We interviewed firms of all sizes in a sample of thirty-five that included the most important Canadian- and foreign-owned corporations in Canada. Approximately one-third of the sampled firms were large (more than 1000 employees), one-third were medium-sized (100 to 999 employees), and the rest were small (1 to 99 employees). More than half the firms (nineteen firms, or 54 percent of the sample) were producers of telecommunications equipment, and the others were manufacturers of semiconductors, computers, and other electronic equipment. We also interviewed two service firms that used a lot of telecommunications or other electronic equipment and that were linked to the manufacturers by technological alliances.

Table 3
Collaborations by electronics firms

Number of alliances	Frequency[a]	Percentage
1 to 4	10	32
5 to 9	8	26
10 or more	13	42
Total	31	100

[a] Missing frequencies (no response) = 4.

The Alliances

Product development was by far the commonest goal: almost half the technical alliances (17 out of 35, or 49 percent) were created for this reason. Other very different goals, such as setting common standards or doing basic research, were far less important. Mixed objectives involving basic and applied research, the setting of standards, and product development were identified in 31 percent of the cases.

Most technical alliances were organized through a simple memorandum of understanding (MOU) that covered the allocated resources, goals, and the time frame of the collaboration. Only ten cases (29 percent) involved the creation of joint ventures or new subsidiaries for R&D. This finding is consistent with a strategy of minimizing transaction costs within alliances, since the forms of organization created by MOUS are more flexible and less costly than joint ventures.

The companies in our sample had been involved, on average, in approximately seven collaborations. But again, the standard deviation was very high: thirteen of them (42 percent) had been involved in ten alliances or more, eight (26 percent) were – or had been – involved in five to nine, while ten (32 percent) had been involved in one to four (Table 3). Most alliances (21, or 62 percent) in our in-depth case studies consisted of only two partners. Among the thirteen other alliances, eight (24 percent) consisted of three to ten partners and only five (15 percent) involved more than ten members (Table 4).

Government funding of technical collaborations was widespread in our survey. A majority of the firms (21, or 60 percent) declared that at least some of the alliances' budgets came from government funds. However, other government resources were seldom solicited: only seven firms had used government research personnel – usually from the National Research Council (NRC) – or other public resources. Moreover, fourteen firms (40 percent) responded that collaborative

Table 4
Number of partners in electronics alliances

Number	Frequency[a]	Percentage
2	21	62
3 to 10	8	24
More than ten	5	15
Total	34	100

[a] Missing frequency = 1.

R&D was financed entirely with the companies' own internal funds, and it is important to note that only one firm was involved exclusively in (three) government-funded alliances. *Technical alliances are thus not government creations but new forms for organizing R&D* in response to dramatic changes in world competition and the new dynamics of accelerated technical change. In Canada, government programs are probably facilitating the trend, but they are not launching it.

Universities were involved in only one third (13, or 37 percent) of the alliances we studied in depth. Usually, academics were solicited as researchers (in eleven cases), but in two cases they were solicited as consultants to an existing alliance. University researchers were thus often solicited to work on the most fundamental aspects of the collaborative projects, and they were often involved in the development of new and improved software.

Most of the alliances in our case studies were new: 79 percent had been formed in 1988 or later; only 21 percent had been formed before 1988. We see the same pattern when we consider all the alliances mentioned by the companies: 64 percent of them were formed in 1988 or after and only 36 percent before 1988. These findings tend to confirm that the wave of technical alliances in the Canadian communication and information industries is a new one, following European and American trends by three or four years.

The main types of management arrangements emerged from this survey. The most frequent (14 cases, or 40 percent) was one in which a dominant member or leader (usually a well-known technical leader in the field) was the manager. The second most common arrangement (12 cases, or 34 percent) was one in which all decisions were taken collectively, usually by a coordinating team in which all member firms were equally represented. Finally, in a few cases (9 cases, or 26 percent) the firms simply coordinated some of their R&D activities, without further organization or complex legal agreement (Table 5).

Table 5
Managing the electronics alliance

Type of management	Frequency	Percentage
Dominant member	14	40
Coordinating committee	12	34
Independent R&D	6	17
Other and mixed types	3	9
Total	35	100

Table 6
Solving the intellectual-property dilemma in electronics alliances

Solution	Frequency[a]	Percentage
Collective property of all members	14	41
Each member owner of its R&D results	7	21
Results belong to the leader	6	18
Results belong to users, suppliers	3	9
Government is the owner	2	6
Other and mixed forms	2	6
Total	34	100

[a] Missing frequency = 1.

Intellectual-Property Strategies

The intellectual-property dilemma could create significant transaction costs if not properly managed: the partners found several solutions (Table 6). The most common one was to decide from the start that intellectual property stemming from the collaboration was to be the collective property of all members of the alliance; this solution was adopted by 41 percent of the partnerships. The second preferred solution was to arrange that each partner would become the owner of the R&D results of its own section of the collective project (21 percent). But in 18 percent of the cases, the results were appropriated by the leader. Other, less frequent solutions to the dilemma are listed in Table 6.

Patents and copyrights for chip design and software had been asked for by one-third of the interviewed companies, either during or after the research process, but three other companies considered that there were still no patentable results. One service firm granted to its foreign manufacturing partner the right to patent the results or to look for

Table 7
Explaining the choice of partner in electronics alliances

Reason	Frequency[a]	Percentage
Technical capacity	24	71
Technical complementarity	20	59
Previous knowledge	14	41
Other reasons	11	32
Size	0	0

Note: Multiple choice question.
[a] Missing frequency = 1.

other types of legal protection. All in all, patent and copyright protection (obtained, in process, or expected) covered a maximum of 50 percent of the cases. In the other cases, either fundamental, non-patentable knowledge was produced, or the partners preferred secrecy as the means of appropriation. For nearly one-third of the alliances (8 cases out of 27) the collaboration was open, without time limits, but specific technical and commercial goals were included. However, the vast majority of the alliances did have a precise time frame that varied from six months to five years. For this group of collaborations a three-year framework was the most frequent.

Technology factors were crucial in a firm's choice of partners (Table 7): the technical capabilities of the prospective partner were most important; the complementary (mainly technological) assets of the partner were slightly less important, and the previous knowledge of the partner was the third most important consideration, showing that trust was crucial in building these close collaborations.

Advantages and Difficulties

Their hopes for opening windows on new technologies, gaining complementary knowledge, capturing new markets, and reducing R&D costs were the most important *ex ante* considerations leading companies to choose alliances as a way of doing R&D (Table 8). Those hopes were already partially realized (Table 9). The most important benefits that the companies could already measure as stemming from the alliance were in the areas of new products (63 percent of the respondents), complementary knowledge (60 percent), increased speed of innovation (49 percent), and R&D diversification (46 percent). However, alliances were considered by the firms in our study to be more important in the long term than in the short term (Table 10), even in this industry where new products are frequently being launched on

Table 8
Reasons for technical alliances in electronics

Reason	Frequency	Percentage
Opening windows on new technologies	18	51
Gaining complementary knowledge	16	46
New markets	15	43
R&D cost reduction	11	31
Demand from users, suppliers	9	26
Other reasons	9	26

Note: Multiple choice question.

Table 9
Advantages of technical alliances in electronics

Advantage	Frequency	Percentage
New products	22	63
Complementary knowledge	21	60
Increased speed of innovation	17	49
R&D diversification	16	46
New clients, suppliers	11	31
Other and mixed responses	7	20
Financing	5	14
Patents	3	9

Note: Multiple choice question.

Table 10
Assessing the results of R&D alliances in electronics

	Short-term results		Long-term results	
Level of importance	Frequency[a]	Percentage	Frequency[b]	Percentage
Not important (0–5)	12	39	7	24
Important (6–7)	5	16	10	34
Very important (8–10)	14	45	12	41

[a] Missing frequencies = 4.
[b] Missing frequencies = 6.

the market. As expected (Winter 1989) companies did not consider patents to be an important benefit in the electronics industry.

Although no firm reported the loss of some measurable benefit as a result of the alliance, there were other difficulties that affected almost 50 percent of the companies during the negotiation of the agreement. By far the most important problem (for 9, or 25 percent

Table 11
Firm size and number of collaborations in electronics

	Frequencies[a] (number of firms)	
Number or collaborations	SME	Large enterprises
Two	15	3
Three or more	6	7
Total	21	10

[a] Missing frequencies = 4.
Note: Chi-square probability = 0.029.
    Fisher's Exact Test (2 tail) probability = 0.036.

of the companies) was the division of the intellectual property. Determining the financial contribution of the members was the second most important obstacle (for 5 companies, or 14 percent). One-third (13 of 35) of the companies also reported a wide variety of difficulties during the implementation of the agreement, including disputes over the ownership of the R&D results (three cases), problems arising from the long-term, nonmarket approach of the university partners (three cases), and bureaucratic problems with public financing (two cases). However, all the reported difficulties were solved through renewed negotiations.

Key Variables

Size was a key variable explaining the collaborative behaviour of the firms. While the SMEs were typically involved in one or two alliances, the large firms were involved in a far greater number (Table 11). For the whole sample the average was seven collaborations per firm. The SMEs devoted a higher percentage of their R&D budget to technical collaborations (Table 12): the unweighted average of collaborative R&D expenditures as a percentage of total R&D expenditures was 10 percent for the large corporations and 27 percent for the SMEs. But large firms, as one would expect, invested larger amounts of money in technical alliances (Table 13). Typically, the SMEs contributed less than Can$1.5 million to these collaborations, while larger firms contributed more than that amount, the average contribution being Can$4 million over the span of the alliance. Nevertheless, these figures tend to confirm that smaller firms find economies of scale in collaborative R&D: with limited resources, SMEs tend to form a smaller number of alliances, but these involve a larger share of their R&D effort.

Table 12
Firm size and R&D collaborations as a percentage of total R&D expenditures
in electronics

	Frequencies[a]	
Collaborations as a percentage of R&D expenditures	SME	Large enterprises
1 to 9.9	5	9
10 or more	16	3
Total	21	12

a Missing frequencies = 2.
Note: Chi-square probability = 0.004.
   Fisher's Exact Test (2 tail) probability = 0.006.

Table 13
Size of firms and average contribution to alliances in electronics

	Frequencies[a]	
Contributions to alliances	SME	Large enterprises
Less than $Can 1.49 million	14	3
More than $Can 1.5 million	2	6
Total	16	9

a Missing frequencies = 10.
Note: Chi-square probability = 0.005.
   Fisher's Exact Test (2 tail) probability = 0.010.

Industry membership was another important variable explaining the alliance behaviour of the firms. Typically, the telecommunications companies were younger firms making more collaborative agreements and seldom looking for patents. The typical goal of their collaboration was the development of products and processes (Table 14). Joint ventures, rather than simple memorandums of agreement, were used to formalize telecommunications-equipment alliances, which focus on product development rather than on research (Table 15).

While most agreements were made within Canadian borders, the trend is clearly towards an increasing number of international alliances. Companies were asked a few questions about all the alliances they had joined. Roughly 40 percent of the 250 partnerships undertaken by the Canadian electronics firms in our sample involved at least

Table 14
Goals of alliances in electronics, by industry

	Industry	
Goal of alliance	Telecommunications	Other
Development	13	4
Other	6	12
Total	19	16

Note: Chi-square probability = 0.010.
　Fisher's Exact Test (2 tail) probability = 0.012.

Table 15
Legal arrangements for collaboration in electronics, by industry

	Industry	
Form of collaboration	Telecommunications	Other
Joint venture	7	1
Memorandum of agreement	12	15
Total	19	16

Note: Chi-square probability = 0.032.
　Fisher's Exact Test (2 tail) probability = 0.037.

one foreign partner. The European Economic Community was the preferred home country of partners; the United States was in second place. In a few cases, Canadian firms were involved in well-known American consortia (for example, MCC or SRC) or in European programs (EUREKA, for example) (Peck 1986; Braillard and Demant 1991). More often, however, Canadian firms of all sizes were forming alliances by themselves, company to company, as they searched for complementarities and highly skilled technical partners. Distance and transaction costs were not an important obstacle to international collaborations in R&D.

There were difficulties associated with government financial participation in technical alliances (Table 16). Because government-financed consortia had more members, intellectual-property sharing was more difficult, and negotiations were longer. On the other hand, since privately financed alliances generally had only two partners, the terms of the agrements were more easily settled.

Table 16
Sources of funding and difficulties during the electronics alliance

Difficulties during the alliance	Origins of funds (frequencies)[a]	
	Private	Public and mixed
Yes	2	11
No	8	7
Total	10	18

[a] Frequencies missing = 7.
Note: Chi-square probability = 0.037.
    Fisher's Exact Test (2 tail) probability = 0.043.

Table 17
Geographic distribution of partners in electronics alliances

Scope	Frequency	Percentage
National	68	45
International	59	39
Regional	25	16
Total	152	100

Regional, National, and International Alliances

Canadian electronics corporations were the most international of the four groups of corporations we studied. Nearly 40 percent of the alliances for which we collected precise data took place in an international setting (Table 17). Most of the international partnerships were conducted by the most competitive sectors of the industry: telecommunications-equipment producers, designers of custom semiconductors, and manufacturers of computer peripherals and flight simulators. International alliances featured European (45 percent) or American (24 percent) partners, with the balance being Japanese and Korean collaborations (15 percent each). This pattern probably emerges because Western European technology is more complementary to Canadian technology in the area of telecommunications equipment, semiconductors, and robotics, and also because EUREKA and the leading Canadian federal programs facilitate Canadian involvement in European alliances. Furthermore, there are probably many more technological partnerships in the EC than there are in the United States, so the chances – and the more established European routines – also favour Western Europeans as partners for Canadian firms.

Table 18
Differences in international and regional alliances in electronics

	Scope	
Partners	International	Regional
Firms	56	5
Universities	3	20
Total	59	25

Note: Fisher's Exact Test (2 tail) probability = 0.000.
   Phi = 0.90.

Few alliances (only 16 percent) were conducted within the regions where the responding company was located, and most alliances involved the responding company with at least one university.[1] Practically no government laboratory was mentioned, but several large provincial government enterprises were very active as users in nationwide or province-wide electronics alliances. National alliances were the most common and often included private electronics manufacturers, crown corporations, universities, and some industry-university research centres. The PRECARN and Vision 2000 consortia, which we shall analyse in the case studies, are in this category.

Mowery and Rosenberg's (1989) observations on the contrasting characteristics of regional, national, and international alliances were entirely confirmed by our study (Table 18). International alliances were completely different from regionally based ones. International alliances were company-to-company research agreements, almost exclusively privately funded, centred on "development" instead of pure research, with large budgets and long-term schedules. Conversely, regional collaborations were industry-university agreements with small budgets and short-term horizons. Basic research was their main goal.

When organizing alliances, companies were looking for complementary technology. Their search was regionally based when the

1 Regions were operationally defined using telephone area codes. Thus, for example, all companies, universities, or public laboratories with area code 416 were put in the Toronto region, all those with area code were put in the Ottawa region, and all those with area code 514 were put in the Montreal region.

Table 19
Differences in electronics alliance goals, by industry

Industry	Goal of collaboration		
	Development	Other	Total
Telecommunications	13	6	19
Other	4	12	16
Total	17	18	35

Note: Fisher's Exact Test (2 tail) probability = 0.012.
  Phi = 0.433.

knowledge was of a more basic or fundamental nature (for example, when they were studying composite materials for the production of custom-made semiconductors). On the other hand, the same corporations scanned the world to organize technical cooperation when the goal was R&D close to production (for example, when they were seeking end-users for a specific new semiconductor).

Industrial Differences in Technical Alliances

Telecommunications-equipment manufacturers were among the most active in collaborations. The majority of them were involved in more than three partnerships (for the others, the mode was only two), and they were more often members of international alliances. In fact, they almost all had some foreign partner. Compared to semiconductor and other electronic equipment manufacturers, telecommunications producers were also exporting more goods and had more extended international operations. The goal of their alliances was more often the development of products (mobile telecommunications, satellites, LAN systems, and other advanced software) than precompetitive research (Table 19). They chose their partners more often on the basis of technical competence as users or suppliers and less on the basis of pure, complementary knowledge (Table 20). Finally, since the development of large and costly systems requires more stable organizational structures, telecommunications-equipment producers were relatively more active in joint-venture agreements than in pure memorandums of understanding (Table 21). On the whole, thus, and in spite of a few major exceptions in other areas, this segment of the industry is the most market-oriented and the most internationalized of the whole Canadian electronics industry.

Table 20
Reasons for choice of partners in electronics, by industry

| Industry | Reason for choice of partner (frequencies)[a] | | |
	Complementary knowledge	Other	Total
Telecommunications	7	11	18
Other	13	3	16
Total	20	14	34

[a] Missing frequency = 1.
Note: Fisher's Exact Test (2 tail) probability = 0.017.
    Phi = 0.43.

Table 21
Differences in electronics agreements, by industry

| Type of agreement | Industry | | |
	Telecommunications	Others	Total
Joint venture	7	1	8
Other	12	15	27
Total	19	16	35

Note: Fisher's Exact Test (2 tail) probability = 0.032.
    Phi = 0.71.

CASE STUDIES OF CANADIAN ALLIANCES

From among the hundreds of cases of technical collaboration in Canadian electronics industries, we selected three to illustrate national and international alliances, both vertical and horizontal. Since all of them have been analysed in detail in the technical and financial press, this study does not reveal any confidential information.

Vision 2000

Vision 2000 is a privately organized, privately and publicly funded research consortium that was created in 1989 with the help of two federal government departments. Its closest foreign model is MCC in the United States (see the next section, below). Its thirty-eight founding members included all the large domestic telecommunications utilities, among them Bell Canada, Alberta Government Telephones,

Telesat, Telecom, and Teleglobe Canada, and also the largest telecom-munications-equipment producers, including Northern Telecom, Mitel, Newbridge Networks, and Spar Aerospace. In May 1991, almost two years after the consortium was created, the first fifteen cooperative research projects were unveiled. The total cost of the research, which focused on communication and information technologies, was set at $30 million. The total cost of the consortium's projects over ten years is evaluated at $1 billion. Half the money comes from the industrial partners and half from the government. The consortium's goals are precompetitive R&D, product and service development, and applied research and standards development in the area of personal commu-nications. The consortium is managed by an executive committee of six members (five from industry and one from the federal Department of Communications).

A major purpose of the partnership is to bring together end users, manufacturers, universities, and public laboratories who are working on specific projects. Nearly 200 persons from these various constitu-encies sit on the advisory boards of Vision 2000: project proposals are selected by the Project Review Board; the R&D Advisory Board con-ducts annual reviews of the research that has been done and estab-lishes themes and priorities; the Markets Advisory Board forecasts market trends; and the Policy and Standards Board provides informa-tion and advice on regulations and standards.

Vision 2000 has no specific rules about intellectual property and leaves each "subconsortium" free to determine its own intellectual-property rules. The consortium as a whole is in fact somewhat like a broker among the members (actual and potential) and conducts no R&D by itself.

In March 1994, Vision 2000 was incorporated into another consor-tium, CANARIE (the Canadian Network for the Advancement of Research, Information, and Education). The president of CANARIE, Ted Strain, suggested that Vision 2000 was a "child of the Department of Communications" and that since there were "too many organisa-tions, ... most companies cannot justify belonging to all these groups" (*Research Money*, 30 March 1994, 7). The fate of Vision 2000 illustrates the somewhat artificial nature of some government alliances.

Northern Telecom's Alliances

The Canadian Microelectronics Corporations (CMC) is a nonprofit university-based consortium of government, industry, and academia. Founded in 1984, it brought together twenty-one university partners from across Canada (from the University of British Columbia to

Memorial University of Newfoundland), four government members (including the federal Department of Communications, the NRC, NSERC, and the Department of Regional Industrial Expansion), and eight industrial partners (including the two largest telecommunications equipment producers, Northern Telecom (NorTel) and Mitel, and a few smaller semiconductor manufacturers). NSERC and NorTel were the main founders of CMC, whose primary purpose is to carry out research and training in integrated circuit design. CMC's revenues grew from $4.8 million in 1985 to $5.5 million in 1990, and its budget was expected to double in 1992, thanks to a massive $24.6 million grant from NSERC for the period 1991–95. CMC is also expected to diversify its sources of funds.

NorTel, Canada's largest electronics firm, also belongs to many international alliances, but these are private arrangements between NorTel and foreign partners. For example, NorTel has been an associate and a shareholder in the Microelectronics and Computer Technology Corporation (MCC) since March 1990 and is represented on its board of directors. MCC is one of the earliest, largest, and most widely publicized American consortia (Peck 1986). It is not an industry-university consortium, nor it is officially linked to any government, though it does have government sponsors including ARPA, NASA, and the U.S. Department of Defense (MCC 1991). Founded in 1982, with headquarters in Austin, Texas, MCC is a privately owned and financed, nonprofit, cooperative joint venture of some twenty large American corporations, including Control Data, Digital Equipment, Martin Marietta, 3M, Motorola, National Semiconductor, and Rockwell International. MCC conducts collaborative research in many areas of computer electronics, including advanced computing technology, computer-aided design (CAD), and software technology and interconnection.

All shareholder companies, those who have bought at least one share in MCC and take part in at least one research project, participate in the governing structure. Associate companies, those who share the funding of at least one research project, take piority in using the technology produced, but MCC holds the intellectual-property rights. Shareholders and associates have the first right to license the technology, but MCC may license the R&D results to third parties after shareholders and associates have exercised or declined to use their first right.

In March 1992, NorTel announced a different type of strategic alliance, in conjunction with Motorola, the world leader in cellular telecommunications equipment. The two companies created a joint venture, Motorola NorTel Communications, based in Chicago and controlled equally by each partner. Each brought its technology and its marketing experience to the alliance, but manufacturing remained with each parent. No other details of the structure were published. The short-term goal of the alliance is to develop, service, and market the partners' existing network technology, but a second phase of the alliance will include joint R&D for the new broad-band network technologies.

NorTel announced another international strategic alliance in July 1992, this time with Matra SA of France. Under this agreement, NorTel buys a 20 percent stake in Matra's telecommunications subsidiary, and 5 to 8 percent of the shares of MMB, the holding company that controls Matra. The goal of this alliance is to combine Matra's advanced radio technology with NorTel's switching-equipment technology. Two joint ventures will be created, one in digital telephone technology, the other in public communications networks. Unfortunately, very few details about the financial, manufacturing, and research dimensions of the agreement have been published.

Techware Systems in SRC

International alliances are not restricted to large firms. The Semiconductor Research Corporation (SRC) is a large, nonprofit, cooperative American venture created in 1982 by the Semiconductor Industry Association. Its members include most of the largest American producers of semiconductors: AMD, Harris, Intel, LSI Logic, Motorola, National Semiconductor, RCA, Texas Instruments, and Zilog. By late 1991, SRC had over sixty-five members, including thirty-six U.S. companies and several consortia, federal laboratories, and universities. Its goal is to perform basic research in semiconductor design, manufacturing, and materials. Techware Systems, a small Vancouver firm that has developed an automated system for the application of thin coats to semiconductors, was admitted into SRC. Techware had lobbied SRC for many months to change the admission rules and permit Canadian firms to participate in the American consortium.

Two other important Canadian electronics consortia worth mentionning are PRECARN, founded in 1987 by thirty-four Canadian corporations to conduct precompetitive research across Canada on artificial intelligence and robotics, and the Solid State Optoelectronics Consortium of Canada (SSOC), created in 1987 with the support of five corporations and the National Research Council of Canada. The goal of SSOC is to develop an optoelectronic integrated semiconductor device. An important university-industry electronics research consortium is the Telecommunications Research Institute of Ontario (TRIO), created in 1988 by the Ontario Centers of Excellence Program and the Ottawa-Carleton Research Institute (OCRI).

These cases illustrate the basic difference between the national and international alliances of Canadian firms: larger budgets, specific development goals, and production and marketing agreements abound in the overseas alliances of Canadian firms. But smaller budgets, university collaboration, and precompetitive research are more often found in domestic consortia.

CONCLUSION

The Canadian electronics industry has entered a phase of intense collaboration in R&D, both in Canada and abroad, with the help of government agencies and departments, especially the National Research Council of Canada and the federal department of Industry. At the provincial and regional level, Ontario, Quebec, and Alberta have been the most active in supporting the development of alliances within their borders. But alliances are also spreading across Canadian borders as Canadian firms develop international alliances, mainly in Europe and the United States. The collaborative trend is not simply a government creation; it is embedded in the new patterns of competition in high-technology industries.

The two most important factors affecting the characteristics of R&D alliances were the size and industry membership of the corporations involved. Smaller firms were active in fewer alliances, but these included most of the R&D effort of smaller firms, which indicates that R&D economies of scale offer part of the explanation for the new collaborative trend. However, the search for complementary assets and high technical skills was probably the most important motive for collaboration in Canadian electronics.

Most of the collaboration took place between manufacturers of telecommunications equipment, both because this sector is more developed than the others and because it is, in the main, under Canadian ownership and control. Foreign subsidiaries in Canada collaborate less, and their links are mainly with universities and government laboratories. Larger partnerships (partnerships with a large budget or a large membership) are always organized by Canadian-owned and controlled firms.

The firms that gained advantages from alliances outnumbered those experiencing difficulties. The benefits included launching new products on the market, acquiring complementary skills, accelerating innovation, and diversifying R&D. By far the most important difficulty was the problem of sharing the intellectual property. This problem arose both during the initial negotiations and during the implementation of the agrement. However, the companies devised several solutions, including collective property rights for all R&D results, individual property rights for each company for its own R&D results, and the appropriation of the R&D products by the leaders of the alliance.

Nonequity collaboration is the most common form in the electronics industry. Arm's-length transactions are the rule: there are few cases of joint ventures and few cases in which alliances have led to mergers. While this finding may tell against the importance of transaction costs for the R&D behaviour of Canadian electronics firms, it is doubtful that it can be generalized. First of all, most alliances have been formed only recently, and some of them could lead to outright mergers and acquisitions in the future. Second, in biotechnology and advanced materials, "protective alliances" between large and small firms did in some cases precede mergers and acquisitions of SMEs by their larger partners.

4 Advanced Materials

Like biotechnology and unlike transportation equipment, advanced materials are products of a technology rather than an industry. They are in fact scattered over many different industries. But unlike biotechnology, advanced materials provide an example of markets pulling innovation, rather than of technology pushing it: in the postwar period, the pull came from the demand for new materials in the defense, electronics, and energy industries. In biotechnology, furthermore, radical innovations abound in both products and processes, but in advanced materials, gradual improvements and incremental innovation are of basic importance, and there are few cases of discontinuities and disruptive breakthroughts. Finally, while all the early development in biotechnology was produced by small, dedicated biotechnology firms, new materials appeared both through the diversification of large materials firms (metal refiners, chemical corporations, glass producers, and forest products firms) and through the entry of small innovative firms.

Nevertheless, like biotechnology, advanced materials are the result of a set of postwar innovations in science and technology. Traditional materials (wood, stone, leather, basic metals, glass, cement, oil and gas, vegetable fibres, commodity chemicals) are extracted from natural sources and produced with few alterations – usually only refining – from their original form. But advanced materials are science-intensive products that emerge not from accumulated expertise but from a deep understanding of the structure of matter (see the appendix to this chapter). They are usually either composites of different existing materials, or they are derived from existing materials through a complex process of design and manufacturing. Advanced materials include metal matrix composites, wood composites, light alloys and super alloys, new fibres (optical fibres, for example), advanced ceramics, and new resins and plastics (Hondros 1986). Most of them were developed in response to demand from the nuclear, electronics, telecommunications, energy, and aerospace industries. Two important events in the history of new materials that revolutionized telecommunications were the development of the first commercial superconductive niobium-titanium wire by Westinghouse in 1962 and the patent obtained in 1972 by Corning Glass in the United States for fibre optics (Diebold 1990). It was also in 1972 that John Bardeen received the Novel Prize in physics for the first successful theory of how superconductive materials work.

Traditional materials cater most often to traditional industries, such as construction, textiles, shoe manufacturing, furniture, fabricated metals, pulp and paper, and the like. Advanced materials are sold mainly to the new industries producing semiconductors and telecommunications and transportation equipment (mainly automobile and aerospace equipment). They also have a special appeal for energy producers looking for materials that can store large amounts of energy or conduct electricity with a minimum loss. Also, they have the potential (and have in fact started) to revolutionize traditional industries like textiles, building construction, and electrical equipment and appliances.

Since advanced materials, unlike traditional ones, are designed from the start to serve specific users, the interaction between users and producers is critical to the development of the materials. Moreover, the scientific characterization of the composites, ceramics, alloys, or fibres is crucial to the understanding of their properties. This is why technical alliances for R&D in advanced materials include users, producers, and knowledge-generating units in universities and government laboratories (Forester 1988; Willinger 1989).

In the United States, most of the public R&D effort in advanced materials is concentrated in two major areas: energy-saving materials (mainly superconductive ceramics) and defense-related composites (Stix 1993). Private R&D expenditures are concentrated in advanced electronic ceramics and structural ceramics for automobile engines. In the EU, publicly funded, privately executed research is concentrated in the aerospace industry (Peters, Groenewegen, et al. 1993).

The largest demand for new materials is for advanced ceramics. The world market for advanced ceramics of all kinds was estimated at US$45–50 billion in 1990, but this figure excludes Eastern European markets and the internal production of vertically integrated firms. The next largest demand was for different types of alloys and superalloys (for the defense industry) and for resin-, metal- and wood-based for composites in a variety of applications. Then is also a major global demand for fibre optics, which had a US$1.75 billion world market in 1990.

ALLIANCES IN ADVANCED MATERIALS

Technical collaboration in advanced materials is the result of both private and public initiatives with either defense or civilian goals. National alliances are more frequently precompetitive coalitions of firms, universities, and public laboratories. Examples of national alliances with government support include several industry-university partnerships and defense-related arrangements, like the joint project of Bellcore and the U.S. Department of the Army, which was launched in 1985 to research gallium arsenate crystals.

Superconductive ceramics consortia were formed in the 1980s in all advanced industrial countries, usually with government support. There are several coalitions for gallium arsenate research, polymer composites, and metals alloys. At the end of 1990, there were twenty such consortia registered in the United States under the National Cooperative Research Act of 1984. The main materials researched by these consortia were superconductive ceramics, semiconductor materials (like gallium arsenate and optical fibres), and polymer-based composites (Rhea 1991). A recent world-wide study shows more than 700 agreements in new materials. Most of the agreements are between American, European and Japanese firms and nearly 90 percent of them have been established since 1980 (Hagedoorn and Schakenraad 1991). However, Canadian firms are not mentioned in this study.

Private international initiatives are of two kinds. Precompetitive research is carried out by international associations of materials manufacturers searching for new composites and alloys based on their products and for new applications for these composites and alloys. The International Magnesium Development Corporation (IMD) and the International Lead-Zinc Research Organization (ILZRO) are among the most active in this area. Cooperative development work is conducted both by firms and by coalitions of horizontally linked firms. In the first case, the goal of the collaboration is to share the high costs of the application and commercial testing of the new materials, since their production enjoys large economies of scale. Examples of this type of collaboration are the carbon-fibre consortia formed by four Japanese firms with their American and European counterparts (Gregory

1987). In the second case, we find alliances between large users and producers of materials who are doing collaborative research for specific applications, for example, research on new alloys and composites for applications in aircraft or other transportation equipment.

Examples of government-supported international alliances include the program under the umbrella of EURAM (the European Research on Advanced Materials Program) in the 1980s and some large international research projects like the International Thermonuclear Experimental Reactor project, an international association of national nuclear fusion projects launched in 1990 to conduct research on plasmas.

Canadian Users and Producers of Advanced Materials

Advanced materials producers belong to a wide range of industries: metal matrix composites are usually manufactured by metals producers; fibre-reinforced plastics are often produced by chemical companies; wood composites are manufactured by pulp and paper and forest products firms. A few large Canadian producers of traditional materials have integrated some R&D and, eventually, some manufacturing of new materials into their operations. Other producers are small and medium-sized firms, founded by university graduates without previous manufacturing experience.

Canadian users of advanced materials come from the transportation, construction, and electrical industries; they include electronics manufacturers and aerospace firms. The handful of users in Canada includes large established companies along with some start-ups. The users do R&D and some production of advanced materials. Since the linkage between users and producers is so important, some large users are vertically integrated for the production of new materials. Knowledge-producing units in Canada include federal and provincial government laboratories, crown corporations, private engineering firms, and two- or three-dozen universities and colleges.

Many metal, pulp and paper, and forest products manufacturers and many oil and gas, chemical, and other materials refiners are potential Canadian producers of advanced materials. But, based on the directories of advanced materials associations, we estimate that there are actually no more than fifty producers, including some large firms in the industries just mentioned, and some small, specialty manufacturers. We estimate that there are only a few users, as well, certainly not more than fifty companies. The total Canadian production of advanced materials is certainly small; its value does not exceed $100 million a year.

The major obstacle to the development of Canadian advanced materials industries is the small number of users in critical industries like defense, transportation equipment, electronics, and high-technology construction. The Canadian defense industry is extremely small compared to the American and Western European industries, and most of the transportation equipment industry is under foreign control and performs little R&D in Canada. A secondary obstacle is the ready availability and low cost of traditional materials in Canada.

Because the domestic market is small, and because incremental innovations and user-producer collaborations are critical, Canadian advanced materials producers tend to locate close to their industrial customers. Some of the largest Canadian manufacturers of advanced materials have transferred their R&D and their production to the United States and the EC, where the large markets are. Inco, for example, the world's largest nickel producer, owns Alloys International, which has plants in Huntington, West Virginia, and Hereford, England, and research laboratories in both countries. According to the *Financial Post* (30 March 1993), Inco is one of the world leaders in the high-nickel alloys used in the aerospace, marine, chemical, and energy industries, among others.

Alcan is also actively involved in advanced materials R&D and manufacturing in foreign countries. It produces most of its new alloys and composites in the United States, Europe, and Japan, although it does produce some of its new materials in Canada. In 1990, for example, it opened a new plant in Quebec for the production of Duralcan, a metal matrix composite used in the production of brake rotors for cars, bicycle, frames, drive shafts, and tire studs, even though Duralcan is sold in the United States (Campbell 1990). Alcan's vice president makes it clear, however, that there are problems blocking the Canadian development of many advanced materials.

A special problem in Canada for materials development is that domestic customers are too few. ... [T]hey demand too little to pull effectively on R&D to upgrade the proposed materials that Canada offers to the world markets. One result in our scientific establishments, in governments and industry, is "technology push" R&D, which might come close, but most often fails to hit the commercial target which is always best understood by the demanding customer and the end user. (Wilson 1991)

THE RESEARCH

In our study of advanced materials, we interviewed thirty-six units (thirty private enterprises and six crown corporations and government laboratories), all of which do cooperative research with other units, both Canadian and foreign. Our sample was representative of users and producers and representative geographically, as well.

Characteristics of the Firms

Our respondents were well-established companies and government units: they were, on average, 35 years old. They were also large: nearly 50 percent had more than 1000 employees and another 14 percent were medium-sized organizations with between 100 and 999 employees. All the respondents conducted R&D regularly, but the average cost of R&D varied widely from metal producers, who, on average, spent 2.5 percent of sales on R&D, to specialized advanced-materials producers, who spent, on average, 21 percent of sales on R&D.

The Alliances

The average respondent was involved in seven collaborations, a figure that closely corresponds to the median (Table 22). The typical alliance (almost three quarters of the alliances) involved two partners, but the average number of partners was three, because there were some very large partnerships with many members (Table 23). Development was the goal of most of the collaborations (78 percent), while the rest were pursuing fundamental or applied research or both.

The collaborations were managed most frequently (15 cases, or 42 percent) by coordinating committees. A hierarchical form of management (13 cases, or 36 percent) with a dominant leader was the second most frequent type (Table 24).

Collaborations were done mainly within Canadian borders: only two large firms conducted a majority of their alliances with foreign partners. Many of the domestic alliances involved government laboratories, crown corporations, and universities. A typical alliance involved a user or producer or both, along with a knowledge-producing unit. International alliances, on the other hand, were usually arrangements between private corporations, without any involvement of government laboratories or universities.

In many cases, partners were chosen for their technical capabilities. Another important consideration was the partner's previous knowledge (Table 25). Respondents also revealed that technical alliances were organized in order to reduce R&D costs (in 58 percent of cases), to foster complementarity (58 percent), and to open windows on new technologies (50 percent) (Table 26). The first reason supports the argument that alliances are formed for R&D economies of scale. The

Table 22
Number of collaborations by advanced materials firms

Number	Frequency	Percentage
1 to 4	18	50
5 to 9	15	42
10 or more	3	8
Total	36	100

Table 23
Number of partners in advanced materials alliances

Number	Frequency	Percentage
2	23	64
3	7	19
4	3	8
5 or more	3	8
Total	36	100

Table 24
Managing the advanced materials alliance

Type of management	Frequency	Percentage
Coordinating committee	15	42
Leader manages	13	36
Independent R&D	1	3
Other and mixed forms	7	19
Total	36	100

Table 25
Choosing the advanced materials partner

Reasons[a]	Frequency	Percentage
Technical capacity	26	72
Previous knowledge	13	36
Complementarity	3	8
Other	16	44
Size	0	0

[a] Multiple choice question.

Table 26
Explaining technical alliances in advanced materials

Reasons[a]	Frequency	Percentage
Reduction of R&D costs	21	58
Complementary technology	21	58
Open windows on new technology	18	50
New products on the market	5	14
Other	2	6

[a] Multiple choice question.

search for complementarity included the search for scientific expertise, for specialized testing equipment, and for inventors and manufacturers. Creating new products for the market was also mentioned as a motive.

As for the legal arrangements, joint ventures were rarely formed. Most often, the alliance partners formalized memorandums of agreement that specified the management organization, the contribution of each member, the division of the intellectual property among the partners, and, usually, secrecy or confidentiality clauses. Open-ended agreements (47 percent) and one- or two-year research projects (22 percent) were preferred.

Intellectual-Property Strategies

The intellectual-property rights for the new materials or related processes belonged to the alliance leader in 42 percent of the alliances, to all members of the collaboration in 33 percent of the alliances, and to the user or the supplier in 14 percent of the alliances. Each partner performed a specific R&D task and kept the research results in 6 percent of the cases. In advanced materials – unlike electronics – it appears much more difficult to share the intellectual property among the partners: sole ownership was the solution in more than 50 percent of the cases.

Contrary to the findings for biotechnology, patents are uncommon in advanced materials. Only ten (27 percent) of the firms had patented or were hoping to patent anything. Secrecy is the preferred strategy for appropriating the R&D results. The respondents considered that in advanced materials, "inventing around" is relatively easy. Secrecy was also considered better than patenting for protecting process technology. Another consideration was that fundamental and applied knowledge stemming from the collaboration (for example,

Table 27
Advantages drawn from advanced materials alliances

Advantages[a]	Frequency	Percentage
Complementary technology	20	56
Increased speed of innovation	16	44
Technology transfer	15	42
Financing	14	39
New products on the market	13	36
Diversification	9	25
Other	9	25
New clients	6	17
Patents	4	11

[a] Multiple choice question.

the characterization of advanced materials and computer-simulation models for studying new processes) were classified as "fundamental," and thus not patentable. Finally, many small companies declared they had no means of protecting their innovations in the courts if a large corporation decided to infringe on them. Most large companies did use patents to protect their product innovations, but few small and medium-sized companies applied for any kind of legal protection.

Advantages and Difficulties

The advantages drawn from alliances were many, and they involved, in our study, the absorption of complementary knowledge (56 percent), followed by accelerated innovation, technology transfer, financing, the launching of new products on the market, diversification, and other strategies (Table 27). As we have seen, patents are uncommon in advanced materials and are not therefore seen as important advantages. On the whole, the firms' expectations, represented by the reasons for choosing collaborations in R&D, were later confirmed by the actual advantages drawn from the alliances.

While all the respondents pointed to some positive benefits they had drawn from the alliances, nearly 50 percent of the respondents experienced difficulties during the negotiation (17 out of 36) and 14 percent (5 out of 36) during the implementation of the agreement. The intellectual-property division was the most difficult issue for the partners: 10 of the 17 firms that had starting difficulties and 2 of the 5 that had problems of implementation found intellectual property to be the most complex problem to tackle.

Long-term considerations are much more important for assessing the results of R&D alliances than short-term considerations (Table 28).

Table 28
Assessing the results of R&D alliances in advanced materials

	Short-term		Long-term	
Assessment	Frequency[a]	Percentage	Frequency[b]	Percentage
Not important (0–5)	14	40	8	24
Important (6–7)	6	17	5	14
Very important (8–10)	15	43	21	62
Total	35	100	34	100

[a] Missing frequency = 1.
[b] Missing frequencies = 2.

Table 29
Size of firms and R&D effort in advanced materials

	Size[a]	
R&D as a percentage of sales	SME	Large enterprises
3.5 or more	13	5
Less than 3.5	3	12
Total	16	17

[a] Missing frequencies = 3.
Note: Chi-square probability = 0.003.
Fisher's Exact Test (2 tail) probability = 0.005.

Advanced materials are usually considered a product of the future; the absence of Canadian markets for these materials, the coming recession, and the scarcity of investment funds put most of these projects out of present production schedules for most companies.

Explanatory Variables

Size was one of the main determinants of company behaviour. Small and medium-sized firms invested a larger proportion of their sales in R&D than large firms (Table 29). Large firms also showed a reduced R&D effort in collaboration as a proportion of their R&D expenditures (Table 30). As the size of the firm grows, it tends to do a larger portion of its R&D in-house; this finding, again, confirms the economies-of-scale argument for alliances.

Smaller firms generally prefer to be the leaders in technical alliances and to keep the intellectual-property rights to the R&D results for

Table 30
Size of firms and collaborative R&D in advanced materials

Collaborative R&D effort as a percentage of total R&D expenditures	Size[a]	
	SME	Large enterprises
Less than 25	5	11
25 or more	11	2
Total	16	13

[a] Missing frequencies = 7.
Note: Chi-square probability = 0.007.
    Fisher's Exact Test (2 tail) probability = 0.008.

Table 31
Size of firms and management of the advanced materials alliance

	Size	
Management	SME	Large enterprises
Leader	15	5
Equal power of partners	4	12
Total	19	17

Note: Chi-square probability = 0.003.
    Fisher's Exact Test (2 tail) probability = 0.006.

Table 32
Size of firms and ownership of R&D results in advanced materials alliances

	Size	
Ownership of results	SME	Large enterprises
Leader	11	4
Other	8	13
Total	19	17

Note: Chi-square probability = 0.037.
    Fisher's Exact Test (2 tail) probability = 0.049.

themselves. Usually, their only technical asset is a new material or family of materials, and they see their leadership and individual intellectual-property rights as conditions for the stability of their firms (Tables 31 and 32).

Difficulties in the negotiation of agreements were statistically linked to public financing (Table 33). Because public supporters prefer larger

Table 33
Public financing and difficulties in the negotiation of advanced materials alliances

	Origin of the funds	
Difficulties in the negotiations	Private	Public and mixed
Yes	10	7
No	17	2
Total	27	9

Note: Chi-square probability = 0.034.
  Fisher's Exact Test (2 tail) probability = 0.055.

Table 34
Scope of advanced materials alliances

Scope	Frequency	Percentage
National	140	61
International	63	27
Regional	28	12
Total	231	100

consortia, where interests, capabilities, and goals are more difficult to reconcile, when public financing was involved, negotiations were longer and intellectual-property issues more prominent.

Regional, National, and International Alliances

In advanced materials, national alliances were much more common than either regional or international alliances (Table 34). In fact, few Canadian corporations were active internationally; the large producers of materials like nonferrous metals and forest products were the most prominent among them. University-industry collaboration at the regional, provincial, and interprovincial levels was very important. Some university research centres (for example, the Ontario Centre for Materials Research at Queen's University, the Ecole Polytechnique at the Université de Montréal) attracted interest from companies across Canada, as did some government research centres, like CANMET (for metal composites and new alloys) and FORINTEK (for wood-based composites).

The few international collaborations of advanced materials producers were very significant, however. Large Canadian metals producers collaborated with their foreign clients in the development of new

Table 35
International versus regional advanced materials alliances

	Scope	
Partners	International	Regional
Firms	57	5
Universities[a]	6	23
Total	63	28

[a] Includes 5 cases with public laboratories.
Note: Fisher's Exact Test (2 tail) probability = 0.000.
    Phi = 0.90.

Table 36
Industrial breakdown of advanced materials alliances

Industry	Number of firms	Number of alliances	Alliances per firm
Chemical	10	44	4
Metallurgy	8	79	10
Forest products	7	56	8
Ceramics	5	10	2
Other	5	42	8
Total	35	231	7

alloys and metal matrix composites. A few forest products corporations cooperated with foreign suppliers and rivals. Again, international collaborations were larger, costlier, and more structured arrangements than national or regional alliances (Table 35).

Industrial Distribution of Alliances

More chemical producers were conducting alliances in advanced materials than producers from any other industry (Table 36). Nevertheless, the average number of collaborations was smaller than in metallurgy. In an industry with a high percentage of foreign control, many subsidiaries of foreign firms in the chemical industry underlined the fact that technical collaboration within Canadian borders was a new activity for them. A few domestically controlled SMEs were also active in the production of polymer-based materials (polymeric fibres and chemical coatings and compounds).

The large domestically owned metallurgists were involved in dozens of alliances, mainly in the international arena, in what seemed to be the other Canadian stronghold, metal matrix composites and alloys. A few domestic SMEs were also active in these fields.[1] A group of forest products firms was conducting R&D on wood composites for the building and furnitures industries and paper coatings for the pulp and paper industries. Some of their alliances were international.

Ceramics was the less important group; by international standards it was thoroughly underrepresented. Two large networks were prominent here: the High Temperature Superconductivity Consortium, centred in Quebec, and the Canadian University-Industry Consortium for Advanced Ceramics, centred in Ontario. Finally, other firms engaged in alliances included producers of mixed materials, who were doing concrete-polymer collaborative research, for example, or materials research in a wide variety of areas or research on fibre optics.

CASE STUDIES OF CANADIAN ALLIANCES

Most advanced materials can be classified on the basis of the material that is used as the matrix, or main element: wood, metals, ceramics, polymers, and so on. The following examples of alliances show the range of advanced materials involved and provide some details about the companies in our survey.

MacMillan Bloedel and Parallam

Based in Vancouver, MacMillan Bloedel is the largest forest products company in Canada and one of the largest in the world. Its strategy is one of diversification through the development of new and improved products, and it conducts the most ambitious research program in the Canadian forest products and pulp and paper industry. Its most important research project was a long-term, collaborative R&D project on new materials that began in the early 1970s and concluded in 1991. This project led to the invention, development, testing, and now manufacturing of a new building material called "Parallam."

Parallam is a wood composite made of parallel strips of wood bonded with waterproof glues. The resulting material is three times as strong as the strongest natural wood and is used in the construction

---

1 A study of emerging technologies by the Science Council of Canada found that plastics and new metal alloys were also strong areas of Canadian expertise in new materials (Steed and Tiffin 1986).

industry, where it is shaped into beams up to 19.8 metres long. The total R&D cost for Parallam, including testing, was $150 million (Annett 1991). The research project was conducted by MacMillan Bloedel in collaboration with the German manufacturer of special new numerically controlled presses that use microwave energy to cure the glue and with a Canadian chemical producer who developed the powerful adhesive required to hold the wood fibres together. Once the R&D project was completed, MacMillan Bloedel spent $100 million to build two Parallam plants, one in the American state of Georgia and one in British Columbia.

The WESTAIM Consortium

Sherritt Gordon Mines is one of the most R&D-intensive corporations in Canadian metallurgy. Since its inception, when it did pioneering research, first on lateritic nickel ores and later, on hydro- and power-metallurgy, Sherritt Gordon has been an R&D-oriented challenger of more established companies like Inco and Falconbridge. In 1989, Sherritt Gordon announced the organization of a large consortium called Western Advanced Industrial Materials (WESTAIM).

Managed and led by Sheritt Gordon, a private, Canadian-owned industrial enterprise, WESTAIM is the largest and most ambitious advanced materials consortium in the history of Canadian industry. The goal of the consortium is to produce a wide variety of advanced materials, including metal matrix composites, ceramics, and polymers. The first five co-organizers of the consortium were Sherritt Gordon, the Alberta government, and three federal government agencies: the National Research Council (NRC), Industry, Science, and Technology Canada (ISTC), and the Western Economic Diversification Fund (WED). Alberta was to provide $40 million, the NRC another $10 million, WED $15 to $20 million, and Sherritt Gordon was to match these public funds to produce a total R&D disbursement of $140 million over five years. Industrial allies (users and specialized manufacturers of materials) and universities were expected to enter the agreement in the near future (*Research Money*, 30 August 1989).

Hydro Quebec and the ACEP Battery

Hydro Québec is the second largest producer of electricity in Canada and one of the top ten spenders on R&D. Its main R&D laboratory, founded in 1970, has collaborated with dozens of suppliers, and it has done collaborative research on alternative technologies for the production and storage of electricity. One such technology involves a new mini-battery.

In the late 1970s and early 1980s, a university researcher in Grenoble, France, invented a new polymer that could accumulate electricity at temperatures between −10°C and 100°C. In collaboration with the French firm Elf-Aquitaine, Hydro Québec used this new polymer to develop a new, small battery called ACEP (*accumulateur à électrolyte polymère*). The three partners signed their first cooperative agreement in 1979, but in 1986 Elf-Aquitaine and the French researcher dropped out of the project, so Hydro Québec bought the patents and continued the R&D by itself. Four years later, in 1990, the lithium mini-battery was ready for large-scale market testing and industrial production.

After searching without success for a North American partner, Hydro Québec signed an agreement with Yuasa Battery, a Japanese industrial firm. The two settled on joint venture in which Hydro Québec and Yuasa were to continue to do the R&D, and Yuasa was to open the first industrial plant in Japan in 1993. A second one was to be opened in Quebec in 1995 after extensive testing of the manufacturing process and the market in Japan.

In this example, the invention was based in a university; most of the innovation was conducted by Hydro Québec in cooperation with Elf-Aquitaine, and the market testing, pilot plant, and industrial production were contributed by the Japanese industrial partner. The profits are to be collected by the Hydro Québec-Yuasa joint venture. Thus this example has all the ingredients of technical collaborations in new materials: complementary partners, a flow of ideas from a university to the private sector, and collective access to the results (Hydro Québec 1991, 7). By 1994, Hydro Québec had signed a similar agreement with the United States Advanced Battery Consortium (a partnership of General Motors, Chrysler, and Ford) to develop a large lithium-polymer battery for future electric cars. The Institut de recherche en électricité du Québec (IREQ), Hydro Québec's central research laboratory, was to provide the battery technology; the 3M Corporation, another partner, was to provide polymer films, and Argonne National Laboratories was to provide battery testing.

Our examples also illustrate the differences between domestic and international alliances. Domestic alliances like WESTAIM usually involve more members, are oriented more towards precompetitive research, and enjoy more governmental support. International alliances are centred more on development; they include production clauses, have, on average, larger budgets for each project, and they often include joint venture agreements.

CONCLUSION

Nearly a hundred Canadian companies and thirty universities, government research centres, and crown corporations are doing R&D in advanced materials. Strategic partnerships were developed among Canadian and some foreign corporations beginning in the mid-1980s and early 1990s. The goal was to foster innovation through complementary technical knowledge and, ultimately, to develop new products. Flexible forms of partnerships (open-ended and short-term agreements instead of joint ventures) were preferred in order to reduce organization costs.

In all our case studies, we found advantages flowing from these alliances, including the absorption of complementary knowledge, faster innovation, and faster technology transfer. But there were persistent transaction difficulties: the division of intellectual property was the most important one.

International alliances were less frequent in advanced materials than in electronics, since Canadian advanced materials firms have little domestic demand and consequently little of the domestic market-pull that is so crucial. Most of the international partnerships in advanced materials were conducted by the largest Canadian producers (and some users, as the example of Hydro Québec shows) with foreign clients. Some of the largest Canadian producers have in fact transferred at least part of their advanced materials research and production to other countries, in order to locate themselves as close as possible to their users in the United States, Western Europe, and Japan.

APPENDIX*

What are Advanced Materials?

- Electronic, magnetic, and optical materials, in particular for the microelectronics and information technology industry; ...
- Technical ceramics, i.e. materials that are formed at high temperature (+800°C) from compounds containing, for example, silicon, carbon, oxygen, nitrogen, aluminum, beryllium, titanium, boron;
- Powder metallurgy, which not only applies to powder-producing operations in metals, but also to aggregation and sintering in ceramics;

* From Hagedoorn and Schakenraad (1991, 429–30).

- Fibre-strengthened composite materials made of two or more substances where the properties of the composites are superior to those of the individual components;
- Technical plastics – chemical materials, a number of whose combined properties satisfy particular requirements such as weight, modulus of rigidity, tensile strength, impact strength, melting point, elasticity, chemical inertness, etc.;
- New materials such as special metals and alloys, in particular serigraphic compounds, aluminium-lithium alloys and amorphous metals.

5 Biotechnology

WITH THE COLLABORATION OF
NATHALIE HADE

Modern biotechnology, the most recent of the technologies in this survey, is "a generic technology rather than a sector, ... [it is] the processing of materials by biological agents, biological agents being understood in this context to include microorganisms, cultured cells, and enzymes" (Walsh 1991). Of course biological agents have been used for centuries to create or improve industrial products: brewer's yeast is an example. In the late nineteenth century and the first half of the twentieth century, biological agents were also used to develop several vaccines. However, modern biotechnology, which is rooted in genetic engineering, is definitely a postwar phenomenon. Its origins can be traced to the discovery, in 1955, of the double-helical structure of deoxyribonucleic acid (DNA) by Watson, Crick, and Wilkins, who later won the Nobel Prize in medicine for their work. But the modern biotechnology industry really began with Cohen and Boyer's method of genetic engineering, which involved splitting genes and recombining them in the DNA. Their invention made almost every biotechnology company a licensee of the two universities (Stanford University and the University of California) where the scientists worked.

The second basic step in biotechnology came in 1975 when the 1984 Nobel Prize winners, Milstein of Cambridge, England, and Kohler of Basel, Switzerland, developed a method to produce monoclonal antibodies. Other important steps wre the discovery of DNA sequencing (1976), the synthesis of the first human growth hormone (1979), the development of a gene-synthesizing machine (1981), the creation of the first artificial chromosome (1983) and the first genetically engineered vaccines (1984) (Stent 1993).

Biotechnology is widely diversified throughout the industrial spectrum; pharmaceuticals (vaccines, diagnostics, hormones, tissue activators) are the most important products, followed by agricultural products (new and improved plants and microorganisms), chemicals (food additives, specialty chemicals), and waste treatment and environmental products. In 1984 the u.s. Office for Technology Assessment estimated that 62 percent of biotechnology products were pharmaceuticals, and that estimate is still valid today (OTA 1984). By 1992, approximately two thirds of the American companies were still focusing on therapeutic or diagnostic applications (Coghlan 1993).

There are two different types of new entrants into the industry. First there is a group of small, inventive start-ups, most of which have university origins. There are nearly 1300 in this group, worldwide. Then in the mid-1980s, large established firms that were already operating in pharmaceuticals, chemicals, food and beverages, and pulp and paper began to enter biotechnology. Since few biotechnology products have found their way to the market, most firms are in fact performing R&D services in biotechnology, either for future in-house manufacturing purposes or for contract research (Sercovich and Leopold 1991; Teitelman 1989). By 1992, the total worldwide production of the new biotechnology industry was estimated to be worth US$8.1 billion (Coghlan 1993).

As our discussion shows, the role of science was, and remains, crucial in the development of biotechnology. "Demand-pull" did not play significant role in the key discoveries that laid the basis for this new generic technology. While the transportation equipment and electronics industries have long since taken off from their scientific base and while most of that technology is now appropriated by private firms, biotechnology is still dependent on scientific developments that take place in university or government laboratories. According to Sharp (1989), biotechnology is now in a stage equivalent to that of microelectronics in the 1950s, when the semiconductor had been developed, but its applications were still to come.

There are several pervasive problems in the emergent biotechnology industry. The first is the unprecedented – and underestimated – cost of R&D and clinical testing. Because different organisms react in different ways to a given drug, both basic research and the clinical testing of biotechnology products appear to take much longer and to be much costlier than was originally expected. Thus, dedicated biotechnology companies (DBCs) have large research budgets, small sales, and few commercial products aside from diagnostic kits for health care and some agricultural seeds and plants. According to the U.S. Pharmaceutical Manufacturers Association, the cost of bringing a drug from the laboratory to the customer is now approximately US$230 million, and it takes twelve years, on average, to do so (Coghlan 1993).

The second problem facing the industry is market resistance. Diagnostic kits and vaccines are expensive, and neither physicians nor hospitals nor governments are keen to introduce them. Even when they have valuable products, human-health DBCs suffer from their lack of experience with the institutional lobbying required to create markets for their products (Green 1992). Their problems become even more difficult when environmental groups, both in the United States and in the EC, resist the introduction of genetically engineered plants and animals for human consumption.

Patenting is the third major obstacle for biotechnology. Many countries either do not recognize patents for like forms or would not give legal protection to new proteins or other microorganisms. For example, only since 1980 has the U.S. Patent Office admitted patent claims for the discovery and isolation or modification of a bacterium found in nature. There are two types of U.S. patents for nonhuman material: a patent may be issued for a state of living matter that did not exist before or for a process that does not occur in nature. But ownership claims for human materials (cells, genes, tissues, and the like) are less clear, and they are subject to a great deal of controversy and litigation.

Patent protection from free riders is expensive and may delay production or licensing. Also, because regulations are more stringent, the patenting process for any kind of drug or life-form is increasingly complex and costly and often out of reach for small DBCs. As a result, some of the most successful biotechnology companies have been busy defending their intellectual property from other companies. Others have associated themselves with larger firms that have more experience with patenting, licensing, and litigation.

ALLIANCES IN BIOTECHNOLOGY

Collaborative research between firms, universities, and public laboratories has been a permanent part of the biotechnology industry since its beginnings in the mid-1970s. Collaboration was at first largely informal, not based on signed agreements, in what Kreiner and Schultz (1990) call "the barter economy" of biotechnology: because of their common background, researchers from the private sector, academia, and government traditionally exchanged their knowledge, testing facilities, and specialized materials. Dedicated biotechnology companies, in turn, developed the spin-offs from the university and government laboratories. This informal collaboration, which is still prevalent, later evolved into more formal types of collaboration.

It was probably the stock market crash of October 1987 that reduced the chances of obtaining funds by floating shares and pushed the specialized American biotechnology firms to look for alternative forms of financing. As a result, small and medium-sized DBCs searched for more permanent collaborations with large pharmaceutical or chemical firms. The DBCs were also hoping to facilitate the more rapid exploitation of their product technology and thus accelerate innovation, to generate a more stable flow of revenues, to share the high costs and risks of new product development, and to obtain access to marketing and distribution networks as fundamental and applied research gave way to the more costly (and, for DBCs, relatively less familiar) clinical testing, commercialization, and manufacturing (Forrest and Martin 1992).

In their analysis of the American biotechnology of human drugs, Forrest and Martin identify a "life cycle" or, less ambitiously, an evolutionary pattern of technical alliances. The cycle begins when DBCs and university and public laboratories sign agreements for fundamental and applied research. Then, as the drugs are about to be developed, large corporations and DBCs form technical alliances without equity positions. And finally, cost and risk considerations push larger corporations to take shareholding positions in their smaller partners.

Hybels' (1990) study of American biotechnology shows that large, diversified corporations were actively acquiring equity positions in DBCs. In more than two-thirds of the cases of ties between DBCs and larger firms, the larger partners were American, but they were Western European in 17 percent of the cases and Japanese, in 13 percent. Large Canadian firms were involved in only 1 percent of the ties. Some of the most important ties led to the absorption of the biotechnology firms by the large chemical or pharmaceutical firms: the late-1980s acquisition of the American biotechnology leader, Genentech, by the Swiss-based multinational pharmaceutical corporation Hoffman-Laroche was the first example of this process.

Several years ago, some authors thought these alliances had little chance of survival (Doz 1988; Killing 1983). The international trend, however, seems to suggest that the links between DBCs and large firms will intensify, as will the links between DBCs and university and government partners. Escalating costs, technological complexities, and high risks all point to an increase in R&D collaboration in commercial biotechnology, both in Canada and elsewhere.

Also, biotechnology is producing a major change in the prescription drug industry: DBCs brought a new technological paradigm (combining molecular biology and chemistry), new and closer links between science and industry, and new research areas (Sapienza 1989). Until the mid-1980s, large pharmaceutical and chemical firms ignored the new technology and remained loyal to their classical research methods. However, the entry into commercial biotechnology was difficult to avoid when sales and government approvals showed some measure of success: in 1991, biotechnology products had sales of more than $4 billion in the United States alone, and biotechnology products won five of thirty new product approvals by the u.s. Federal Drug Administration. Even if biotechnology research was slower and costlier than predicted, some results were already on the market. Since they were unable to internalize all the necessary knowledge and research personnel, large pharmaceutical and chemical firms started a process of technical collaboration with DBCs. The trend has only increased since the stock market crash of October 1987 made it more difficult for DCBs to float new shares in North American markets and pushed them to collaborate more frequently. Early in 1992, collaboration received a new impetus when the market value of DBC shares tumbled again with the news that several of the most promising drugs of American companies did not pass clinical tests.

As a result of the soaring costs of the biotechnology of drug development, the largest technical alliance of all was announced in April 1993. Fifteen of the largest American and European multinational pharmaceutical corporations reached an agreement to collaborate in the search for a treatment against HIV, the virus that causes AIDS (*Financial Times* (London), 21 April 1993). The companies were to swap information on clinical trials and development techniques. The American corporations include Merck, Bristol-Myers Squibb, Eli Lilly, Pfizer, Du Pont Merck, and Syntex. The Europeans were Glaxo and Smith-Kline Beecham (UK), Roche (Switzerland), Hoechst and Boehringer Ingelheim (Germany), Astra (Sweden), and Sigma-Tau (Italy).

Canadian Biotechnology

In 1980, according to a Canadian government study, there were 33 companies in Canada conducting biotechnology R&D. Of these, only three were emergent biotechnology firms; the others were established food, beverage, and chemical firms (Ministry of State for Science and Technology 1980). By 1986, there were nearly 100 companies (Province of Ontario 1988), and by 1991, there were 267 companies in Canada directly engaged in biotechnology research. Among the 88 companies for which the *Canadian Biotechnology Directory* (Canadian Biotechnology Association 1991) provided a founding date, only 12 were created before 1970, 13 between 1970 and 1979, and the remaining 53 were created in 1980 or after.

The 267 companies doing research include a majority of DBCS but also a large number of firms using biotechnology for specific industrial purposes. These include agricultural firms, chemical producers, consulting engineers, food and beverage manufacturers, and mineral producers, as well as pharmaceutical, pulp and paper, and waste- and environmental-management firms. A few of them are large established producers, but the majority are DBCS, young, small firms with few employees and low sales. None of these Canadian DBCS has more than 500 employees, except for Connaught Laboratories, which recently became the Canadian subsidiary of the French government corporation Institut Mérieux. The largest private Canadian DBCS include Allelix, Biomira, Biochem, and Quadra Logic Technologies.

It would be difficult to explain the sudden rise of Canadian biotechnology in the 1980s without making explicit reference to government nurturing of this new area of research. In 1983, to begin with, the federal government launched the National Biotechnology Strategy and provided significant funding. Ottawa had recognized Canadian backwardness in this key area and had decided to catch up with other industrial countries through a vast multidimensional effort. The Division of Biotechnological Sciences, part of Canada's main government laboratory (the NRC) in Ottawa was to become the principal public arm for this technology. A $61-million Biotechnology Research Institute (BRI) for industrial research was created in Montreal and inaugurated in 1987. It started working in four areas: generic engineering, protein engineering and cell fusion, biochemical engineering and fermentation processes, and molecular immunology. BRI now owns the most advanced biotechnology pilot plant in Canada and employs nearly 250 scientists and engineers. Another NRC institute, the Saskatoon National Research Laboratory, was upgraded to become the new Plant Biotechnology Institute (PBI), which employed 170 scientists and technicians. The NRC's biology laboratories in Ottawa were also revamped to become the Institute for Biological Sciences (IBS), which centred on cell systems, structural immunology, protein structure and design, and biomedical NMR (nuclear magnetic resonance). By 1991, this new institute employed 250 scientists and other technical personnel.

By 1991–2, these three institutes were the largest Canadian research laboratories in biotechnology. The Institute for Marine Biosciences in Halifax was also created as part of the same strategy. Altogether, these government laboratories employ more than 700 researchers and supporting staff and have combined budgets of more than $60 million. In addition, the Agriculture Canada laboratories and the federal

Forest Research Center have started or increased their involvement in biotechnology projects. Although it is much smaller than the other biotechnology laboratories, Agriculture Canada's research station in Regina employs ten scientists and ten technicians. The federal government has also launched several industry-university research funding programs. And finally, several federal and provincial centres of excellence were funded in the 1980s to link the government laboratories with private firms and universities.

Following the federal initiatives, several provincial governments, including those of Quebec, Saskatchewan, Ontario, and British Columbia, developed their own policies for biotechnology. In 1980, the Manitoba Research Council created a biotechnology laboratory, and the Alberta Research Council followed suit in 1983 with a biotechnology department. In 1985, the Atlantic Institute for Biotechnology was created, with partial funding from Industry, Science and Technology Canada.

The federal and provincial initiatives created the environment in which nearly 200 DBCs were founded. These small firms are conducting R&D, either for larger corporations on a contract basis or in-house, with the ultimate goal of marketing present or future products like diagnostic tests, vaccines, enzymes, or hormones. In 1991, according to a recent survey, 30 percent of Canadian biotechnology products were in the research stage, 35 percent had started trials, and another 35 percent had reached production (Ernst and Young 1992).

The two major Canadian strengths in biotechnology are in health care products, where nearly half of the firms operate close to the pharmaceutical industry, and in agricultural products, where 26 percent operate. In the United States, the distribution is much more skewed: 63 percent of the companies are involved in health care and only 8 percent in agricultural biotechnology (Ernst and Young 1992). Most of the Canadian industry is concentrated in three provinces: Ontario, Quebec, and British Columbia. Toronto, Montreal, and Vancouver are the major centres.

Technological alliances occur between SMEs and larger firms, but there is also a significant amount of collaboration between private enterprises of all sizes, universities, and government laboratories. When SMEs collaborate with larger firms, the SME already owns an invention, typically a partially developed, patented product, and the SME needs the larger firms' funds to fully develop the product through clinical tests. It also needs the larger firms' manufacturing technology and legal know-how, in order to obtain government approval and manufacture the product. When private firms (large or small) collaborate with universities or governments, they seek to acquire fundamental knowledge from the university or government laboratory in order to discover or characterize a product.

In spite of government efforts, Canadian DBCs are small when compared with American corporations like Genentech, Biosystems, Cetus, Centocor, and Amgen, each of which spends more than $40 million annually on R&D and has sales of more than $50 million. Several factors have been mentioned to explain Canadian backwardness: inadequate patent protection and regulations, insufficient financing, a lack of human resources, and poor market acceptance for new products. It has been suggested as well that even if their research is excellent, the capacity of scientists to become entrepreneurs is, nevertheless, still limited (National Biotechnology Committee 1991).

An indication of the current market sentiment about Canadian biotechnology is the recent evolution of the biotechnology and pharmaceutical index on the Toronto Stock Exchange. Although it was launched in February 1992 at 1000, the index was at only 450 on 25 May 1994, and only 39 DBCs were quoted on the Canadian stock markets. The difficulties that the DBCs have experienced in finding R&D funds through the capital market has been a major inducement for their alliances: through private arrangements they try to counteract the indifference of the open market.

THE RESEARCH

Characteristics of the Firms

In our study we conducted interviews with 35 companies of different sizes (nine were large companies) across Canada in the areas of pharmaceuticals (63 percent), environmental products (11 percent), food products (9 percent), agricultural and forestry products (6 percent), and other or diversified products (11 percent). Fully two-thirds of the companies in our sample were young firms founded in 1981 or after; nevertheless, the average age of the firms in the sample was 23 years, because a number of older food and beverage, pulp and paper, chemical, and mining firms were active in biotechnology R&D.

The Alliances

Development was the most important goal of the alliances (68 percent, followed by mixed R&D objectives, usually fundamental and applied research (26 percent), or pure research (6 percent). Most firms were involved in several partnerships at the same time: each firm had made

Table 37
Number of collaborations per firm in biotechnology

Number	Frequency	Percentage
1 to 4	16	46
5 to 9	12	34
10 or more	7	20
Total	35	100

Table 38
Number of partners in biotechnology alliances

Number	Frequency[a]	Percentage
2	22	65
3 to 10	12	35
More than 10	0	0
Total	34	100

[a] Missing frequency = 1.

an average of five collaborative agreements. Also, the average number of partners in an alliance was three (Tables 37 and 38).

The alliances were typically medium-term governance structures, because, in the first place, the goals of the alliance, the contributions of the members, and the intellectual-property division of the R&D results were formalized most often through memorandums of agreement (in 80 percent of cases), a less stringent form of organization than joint ventures, which appeared in only a minority of cases (20 percent). Also, as in the other industries, R&D was typically conducted in the companies' own laboratories (in 80 percent of cases), indicating a more flexible strategy and a less ambitious commitment than is indicated by the creation (in 20 percent of cases) of new laboratories for the alliance. Another indicator of the search for flexibility was the duration of the agreements, which lasted three years or less in approximately two-thirds of the cases. Only 29 percent of the companies chose agreements lasting five years or more. Finally, the management scheme was typically the most flexible available: almost two-thirds of the alliances were organized by coordinated, but nonetheless independent, R&D activities of the members (Table 39).

Partners were chosen most often in Canada (53 percent), but some companies indicated a mix of Canadian and foreign partners (44 percent). The technical capability of potential partners, complementary

Table 39
Managing the biotechnology alliance

Type of management	Frequency[a]	Percentage
Dominant member	1	3
Coordinating committee	11	33
Independent R&D	20	61
Other and mixed forms	1	3
Total	33	100

[a] Missing frequencies = 2.

Table 40
Explaining the choice of biotechnology partners

Reason	Frequency[a]	Percentage[b]
Technical capacity	21	62
Complementarity	18	53
Previous knowledge	18	53
Size	–	–
Other	2	6

[a] Missing frequencies = 2.
[b] Multiple choice question.

skills, and previous knowledge were by far the most important considerations in picking a partner (Table 40). The alliances were almost equally split between those with government and private funding and those with only private funding. This finding confirms that obtaining government research funds was not the main goal of the alliances. Related to this is the finding that strategic alliances were chosen as an instrument for technical development by firms searching for new markets (52 percent of the firms), open windows on new technologies (39 percent), a reduction in R&D costs (33 percent), demands from users or suppliers (33 percent), or complementarities (29 percent) (Table 41).

Intellectual-Property Strategies

Ninety-two percent of the alliances (33 cases) had obtained research results, and the intellectual property was typically shared collectively by the partners. This finding is consistent with the absence of strong leaders, and also with the fact that in almost half the cases the output was fundamental and thus not patentable knowledge. Collective ownership of results is easier when the output is fundamental knowledge.

Table 41
Reasons for alliances in biotechnology

Reason	Frequency[a]	Percentage[b]
New markets	17	52
Open windows on new technology	13	39
Demands from users, suppliers	11	33
Reduction of R&D costs	9	27
Complementarities	9	27
Other	11	33

[a] Missing frequencies = 2.
[b] Multiple choice question.

Of the 19 alliances that did have patentable results, almost two-thirds had patented them. The patents covered products much more often than processes; when the output was a process, secrecy was frequently used to keep the knowledge within the consortium. In general, patents are the key means of appropriating biotechnology. This finding confirms Winter's (1989) conclusion about the industry-specific usefulness of patents as a means of appropriating technology.

Advantages and Difficulties

The advantages of alliances outnumbered the difficulties, and all firms but one found benefits from their alliances. Among the most frequent advantages mentioned were the absorption of complementary knowledge (by 71 percent of the firms), the possibility of launching new products onto the market (62 percent), an increase in the speed of innovation (53 percent), and a diversification of R&D (47 percent). In contrast to the pattern for new materials and electronics, patents were mentioned by 41 percent of the firms as being part of the benefits (Table 42).

A majority (62 percent) of the firms identified no specific difficulty, thus bringing some additional evidence against the usefulness of the transaction-cost approach to analysing alliances. As with the other advanced technologies, the most frequent difficulty (4 cases) in negotiating alliances in biotechnology was the sharing of intellectual property. Other difficulties were also important: seven firms did make some references to transaction costs: namely, the obstacles they met in coordinating different corporate cultures and languages and in dealing with partnerships as a new type of R&D organization. Three companies mentioned difficulties related to the financial terms of the agreement.

Table 42
Advantages of biotechnology alliances

Advantage	Frequency[a]	Percentage[b]
Complementary knowledge	24	71
New products on the market	21	62
Increased speed of innovation	18	53
R&D diversification	16	47
Patents	14	41
New clients, suppliers	10	29
Financing	10	29
Other	6	18

[a] Missing frequency = 1.
[b] Multiple choice question.

Table 43
Assessing the results of R&D alliances in biotechnology

	Short term		Long term	
Assessment	Frequency[a]	Percentage	Frequency[a]	Percentage
Not important (0–5)	15	44	6	18
Important (6–7)	3	9	8	24
Very important (8–10)	16	47	20	58
Total	34	100	34	100

[a] Missing frequency = 1.

Are alliances important? They seem to be less important in the short term. Whatever their present benefits, companies expect larger and more important returns in the long term than in the short term (Table 43).

Explanatory Variables

Size and age are among the most important variables that influence the behaviour of firms doing research in biotechnology. Both variables divide the population into two groups: the small, specialized biotechnology firms are in one group; and in the other group are the large pharmaceutical, chemical, food, and pulp and paper firms that are doing some research in biotechnology, together with some R&D in other areas.

Size was a key factor explaining the firms' R&D effort (as measured by total R&D expenditures as a percentage of sales). As in other

Table 44
Size of biotechnology firms and R&D effort

R&D as a percentage of sales	Size[a]	
	Large enterprises	SMES
Less than 55	8	11
55 or more	0	13
Total	8	24

[a] Missing frequencies = 3.
Note: Chi-square probability = 0.007.
    Fisher's Exact Test (2 tail) probability = 0.01.

Table 45
Size of biotechnology firms and R&D personnel

R&D personnel as a percentage of total personnel	Large enterprises	SMES
44 or less	9	8
More than 44	0	17
Total[a]	9	25

[a] Missing frequency = 1.
Note: Chi-square probability = 0.000.
    Fisher's Exact Test (2 tail) probability = 0.001.

technologies, smaller firms are engaged in a stronger R&D effort than larger ones (Tables 44 and 45), since entry into an industry with technical barriers commands heavier research and development costs for newcomers.

The size of a firm is linked to its age: younger firms (typically the biotechnology start-ups) are smaller than older ones (Table 46). Pharmaceutical research and products (diagnostics kits, vaccines, and the like) are the most frequent outputs of those younger SMES (Table 47). Younger firms also have lower sales, but a higher ratio of R&D to sales (Tables 48 and 49). Finally, their alliance activity is also the most recent (Table 50).

Once again, as in the other samples, difficulties were statistically linked to public financing (Table 51). Alliances with public funds had more members, were more focused on precompetitive activities, and suffered from longer negotiations.

Table 46
Age and size of biotechnology firms

	Age (years)[a]	
Size	Less than 15	15 or more
Large enterprises	1	8
SME	21	4
Total	22	12

[a] Missing frequency = 1.

Note: Chi-square probability = 0.000.
    Fisher's Exact Test (2 tail) probability = 0.000.

Table 47
Age and products of biotechnology firms

	Age (years)[a]	
Products	Less than 15	15 or more
Pharmaceutical	17	5
Other	5	7
Total	22	12

[a] Missing frequency = 1.

Note: Chi-square probability = 0.038.
    Fisher's Exact Test (2 tail) probability = 0.060.

Table 48
Age and sales of biotechnology firms

	Age (years)[a]	
Sales (Can $ million)	Less than 15	15 or more
Less than 11	17	1
11 or more	3	10
Total	21	11

[a] Missing frequencies = 4.

Note: Chi-square probability = 0.000.
    Fisher's Exact Test (2 tail) probability = 0.000.

Table 49
Age of biotechnology firms and R&D effort

R&D as a percentage of sales	Age (years)[a]	
	Less than 15	15 or more
Less than 55	9	10
55 or more	12	11
Total	21	21

[a] Missing frequencies = 3.
Note: Chi-square probability = 0.024.
   Fisher's Exact Test (2 tail) probability = 0.011.

Table 50
Age of biotechnology firms and age of alliances

Age of alliance (years)	Age of firm (years)[a]	
	Less than 15	15 or more
2 or less	12	2
More than 2	9	10
Total	21	12

[a] Missing frequencies = 2.
Note: Chi-square probability = 0.024.
   Fisher's Exact Test (2 tail) probability = 0.033.

Table 51
Difficulties in biotechnology alliances and source of funding

Difficulties in the negotiation	Source of R&D funds[a]	
	Private	Public and mixed
No	6	7
Yes	3	18
Total	9	25

[a] Missing frequency = 1.
Note: Chi-square probability = 0.041.
   Fisher's Exact Test (2 tail) probability = 0.057.

Table 52
Geographic scope of biotechnology alliances

Scope	Frequency	Percentage
National	186	76
International	22	9
Regional	36	15
Total	244	100

Regional, National, and International Alliances

Industry-university collaboration was widespread in biotechnology, as was the cooperation of industry with public laboratories and research hospitals. These types of collaboration were either regional or, more frequently, national in scope (Table 52). The biotechnology laboratories of the NRC and Agriculture Canada were the most often solicited public laboratories.

Dedicated biotechnology companies also collaborated with other corporations, both Canadian and foreign, in the development of products and processes. As in previous samples, regional collaboration usually involved universities, but in contrast to previous samples, international collaboration was equally divided between company-to-company and industry-university collaborations. Canadian firms collaborated with foreign corporations in the clinical testing and final development of drugs, seeds, and other biotechnology products. These overseas collaborations were the most costly and structured of all alliances in the industry; the Biochem Pharma and the Quadra Logic stories in this chapter will illustrate this kind of arrangement. International allies most often brought manufacturing and marketing expertise to the small Canadian DBC.

Foreign universities were sometimes enrolled in R&D projects because of their competence in specific fields. However, most industry-university collaborations in Canadian biotechnology did not cross national borders (Table 53).

Areas of Research

In Canada, as in the United States, most of the research effort in biotechnology is concentrated in the biotechnology of human health, since this technology is supposed to have a larger potential market. Our sample included a majority of firms active in this area. Firms involved in the biotechnology of human health differed in a number

Table 53
Membership of international and regional biotechnology alliances

	Scope	
Partners	International	Regional
Firms	11	6
University and public laboratories	11	30
Total	22	36

Note: Fisher's Exact Test (2 tail) probability = 0.000.
   Phi = 0.90.

Table 54
Age of biotechnology firms, by industry

	Age of firm[a]		
Industry	Less than 15 years	15 years or older	Total
Human health	17	5	22
Other areas	5	7	12
Total	22	12	34

[a] Missing frequency = 1.
Note: Fisher's Exact Test (2 tail) probability = 0.045.
   Phi = 0.356.

of ways from firms in all other sectors. First, they were younger: most
of the human-health DBCs were less than fifteen years old, while the
opposite was true for the larger agriculture and food, pulp and paper,
and engineering companies working in other areas of biotechnology
(Table 54). This finding contrasts the new DBCs (most of which are
working on human vaccines and diagnostics) with corporations from
other industries that perform some R&D in biotechnology. Second,
human-health DBCs had larger budgets for biotechnology R&D, which
is understandable, considering the high cost of biomedical research
compared to research in other areas (Table 55). Also, human-health
DBCs engaged in many more alliances than the other firms, and
networking was much more prevalent in the more complex and
demanding area of health than in waste treatment, agriculture, and
the like (Table 56). In fact, large-budget international alliances exist
only in the area of human health. Finally, we note that biomedical
firms typically demanded patents for their inventions more often than
other firms (Table 57).

Table 55
Differences in biotechnology budgets, by industry

| Industry | Typical alliance budgets[a] | | |
	Less than $Can 0.5 million	$Can 0.5 million or more	Totals
Human health	7	11	18
Other areas	8	1	9
Total	15	12	27

[a] Missing frequencies = 8.
Note: Fisher's Exact Test (2 tail) probability = 0.019.
   Phi = 0.474.

Table 56
Biotechnology collaborations, by industry

| Industry | Number of collaborations per firm | | |
	Less than 5	5 or more	Totals
Human health	7	15	22
Other areas	10	3	13
Total	17	18	35

Note: Fisher's Exact Test (2 tail) probability = 0.06.
   Phi = 0.436.

Table 57
Biotechnology patenting, by industry

| Industry | Patents[a] | | |
	Yes	No	Total
Human health	12	6	18
Other areas	1	10	11
Total	13	18	29

[a] Missing frequencies = 6.
Note: Fisher's Exact Test (2 tail) probability = 0.003.
   Phi = 0.562.

CASE STUDIES

Most of the alliances in biotechnology have taken place in the medical area, where new biotechnology drugs are being researched. The following cases are representative of the major international alliances that are forged in this new section of the pharmaceutical industry.

BioChem Pharma

BioChem Pharma is a medium-sized biotechnology company that started its operations in 1986, when a group of scientists bought the pharmaceutical facilities belonging to the Institut Armand Frappier of the Université du Québec at Laval, north of Montreal. Soon after, IAF BioChem (the original name of the company) floated its first shares on the Montreal Stock Exchange, under the Quebec Stock Savings Plan. The company is currently controlled by its management, composed mainly of scientists, who hold 10 percent of its shares. Two institutional investors, the Quebec Savings and Investment Fund (Caisse de dépôt et de placement du Québec) and the Solidarity Fund of the Quebec Federation of Labour, hold a majority of the company's shares.

BioChem is active in three business sectors: therapeutics, diagnostics, and vaccines. In the first sector, it has developed two promising new drugs. One is 3TC, an anti-AIDS drug. In 1990 3TC was picked by the National Cancer Institute of Washington as the one of the best and most promising candidates to replace AZT, the only product presently used, but one with toxic side effects. 3TC was developed within BioChem, primarily by the late Dr. Bernard Belleau. The second promising drug is BCH-242, an anticancer product that is also less toxic and more effective than any other existing compound. BioChem also produces vaccines, fine chemicals, and diagnostic kits. For the year ending 31 January 1994, the company had revenues of $42.2 million.

Early in 1990, BioChem announced a strategic alliance with Glaxo Holdings PLC, the British-based pharmaceutical multinational. The alliance gave Glaxo exclusive rights to 3TC in all countries except the United States and Canada. Glaxo paid $15 million for those rights and, at same time, agreed to collaborate with BioChem on preclinical research. Later in the year, in November 1990, the two companies extended their agreement to cover the anticancer drug. Glaxo was to form a joint venture for R&D and marketing with BioChem, pay $25 million for a 10 percent equity interest in it, and hold a two-year option for another 10 percent of BioChem's shares. Meanwhile, in 1990, the chances for the success of 3TC went from 25 percent to 50 percent as preclinical R&D progressed. By the end of 1994, the third phrase of the clinical trials was completed, and manufacturing was to start in 1995.

The Biochem-Glaxo alliance is one in which a Canadian SME with a strong R&D base (forty Ph.D.s in a research staff of sixty) but less than 300 employees collaborates with a giant firm with 38,000 employees and an annual R&D budget of more than $1.2 billion. The small firm contributes its innovative products to the alliance, and the large multinational corporation adds its financial, marketing, and technical strengths.

In the field of diagnostics, which is less R&D-intensive and closer to its working alone experience with vaccines, BioChem is working alone. As a result of its acquisition in May 1994 of Ares-Serono, a Swiss company, with revenues of $90 million and manufacturing subsidiaries in the United States, Italy, and Great Britain, BioChem expects to become one of the ten or fifteen largest diagnostics companies in the world.

Allelix Biopharmaceuticals

In November 1988, Allelix Biopharmaceuticals of Toronto and Glaxo Canada announced a strategic alliance for a $10 million R&D project. It was a five-year, renewable agreement to develop ALX1–11, a drug for the treatment of osteoporosis based on recent discoveries in the molecular biology of bone formation and decay. Glaxo is the second largest drug company in Canada; Allelix, with over 75 scientists involved in R&D and total employment of less than 200, is a medium-sized firm, but one of the largest in Canadian biotechnology. Founded in 1983, Allelix is a typical DBC, with revenues of $7.5 million in 1993 and R&D expenditures of $13 million. Allelix owns a portfolio of 18 granted U.S. patents; it has more than 40 U.S. patents pending, and it made 14 new U.S. patent applications in 1994. Although it is less diversified than BioChem Pharma, Allelix has several products under development for the treatment of AIDS, local bone growth factors, and so on.

Glaxo Canada was to supply the R&D funds for the alliance, and Allelix was to contribute its scientific talent and previous research on the use of hormones in the treatment of illness. Glaxo took a 14-percent equity stake in Allelix, but in 1993, when the alliance was renewed, it exchanged this equity for equity in a newly formed joint venture, Glaxo-Allelix, in order to pursue the clinical development of ALX1–11. If the project results in commercially valuable intellectual property, both companies will share the patents, and Glaxo will keep the worldwide marketing rights and Allelix a royalty of between 4 and 8 percent on gross sales of the product (*Research Money*, 16 November 1988).

Allelix is also engaged in national alliances, including a $1.5 million partnership signed in November 1989 with the Biotechnology Research Institute, the NRC's Montreal-based public laboratory. The goal of the applied-research partnership is to develop new drugs for several diseases of the central nervous system, such as migraines, schizophrenia, anxiety, and depression (*Research Money*, 29 November 1989).

Quadra Logic and American Cyanamid

In 1991, Quadra Logic Technologies of Vancouver announced a strategic alliance with American Cyanamid. Based in New Jersey, American Cyanamid is one of the largest U.S. chemical companies. Quadra Logic, on the other hand, is a small biotechnology firm founded in 1981 by a group of scientists headed by Dr Julia Levy at the University of British Columbia. Quadra Logic is active in the area of photodynamic anti-cancer serums and bases its work on a technology it bought from a subsidiary of Johnson and Johnson in the United States. In this major alliance, American Cyanamid paid $20 million to cover 50 percent of the cost of codeveloping a new photodynamic drug for treating venereal warts and psoriasis. In exchange, Cyanamid acquired the exclusive global right to distribute and license future products. Quadra Logic was to contribute the remaining 50 percent of the costs. After the alliance was formed, American Cyanamid took a 15 percent equity position in Quadra Logic (*Financial Post*, 15 May 1992; *Financial Post Magazine*, February 1991). In April 1993, Quadra Logic finally received federal approval to market Photofrin, its laser-actived, cancer-fighting drug. This was the first major drug produced by the new biotechnology firms in Canada. The total R&D cost of the drug, including clinical trials, had been approximately $80 million. Ottawa approved the drug only for the treatment of bladder cancer, but the drug has applications for other diseases as well, including cancers of the lung, stomach, cervix, and esophagus. Approximately 5000 cases of bladder cancer are diagnosed in Canada annually, and the potential market for the drug is nearly three-quarters of that figure (*Financial Post*, 27 April 1993). Later in 1993 and 1994, the United States and the Netherlands approved the drug for the treatment of lung and esophageal cancers; other approvals were expected in 1995.

These case studies reveal the differences between national and international alliances. National alliances are smaller, precompetitive, government-funded agreements. International alliances have larger budgets; they tend to be firm-to-firm deals between private partners with no government participation; they emphasize development and therefore include production and marketing clauses.

CONCLUSION

Biotechnology is the newest generic technology. The Canadian industry is formed mainly by 250 small firms and approximately two dozen large, established corporations operating in the areas of food products, pharmaceuticals, chemical products, mineral refining, and pulp and paper products. Biotechnology is an example of "science-and-technology push," in which universities and government laboratories are still important and will continue to be in the foreseeable future. Some of the most successful companies are, in fact, university spin-offs.

Technical alliances brought several benefits to the companies involved, including financing, complementary knowledge, innovation, patented technology, new products for the market, and new clients and suppliers. As in other areas, biotechnology alliances are medium- and long-term endeavours with many returns expected in the future and some advantages actually realized. Complementary assets are the most evident advantages frequently won by partners in this industry: the most obvious examples occur when small inventive start-ups that have innovative proprietary-product technology but little process technology and few financial resources or marketing skills form alliances with large pharmaceutical companies that do have manufacturing-process technology, large R&D laboratories, clinical testing, patenting, and marketing experience, and ample financial resources.

In contrast to the pattern in electronics, biotechnology firms, mainly in pharmaceuticals, have increasingly used equity clauses in their alliance agreements. This should not necessarily be interpreted as a case of transaction-cost reduction; it may be the normal evolution of a relationship between economic agents possessing perfectly complementary assets.

6 Transportation Equipment

WITH THE COLLABORATION OF
MICHÈLE SAWCHUCK

It may seem odd to include transportation equipment in a study of alliances in advanced technologies. After all, self-propelled transportation equipment has existed for two centuries. Steamships were invented in the late eighteenth and steam locomotives in the early nineteenth century, soon to be joined by the internal combustion engine in 1860 and the electric railway and the aircraft in the early twentieth century (McNeil 1990).

However, there are several reasons for including transportation equipment in our study. First, this is one of the most important industries in which technological alliances take place. Any study of strategic technological partnerships would be incomplete without a chapter on transportation equipment. Second, most of the innovations in two of the three new generic technologies (electronics and new materials) are applied to transportation equipment. Electronic controls and telecommunications equipment have transformed transportation equipment, while robots and other automated machinery have revolutionized its manufacturing processes. New alloys, superplastics, and other new materials are becoming increasingly significant in the industry. The transportation equipment industry is in fact a technological borrower from other, newer industries. Its demand induces technical change in both materials and electronics. Thus, many alliances take place between transportation equipment producers and firms operating in one or several of the new generic technologies.

Third, "just-in-time" system of production and other organizational innovations have been adopted more frequently in the transportation equipment industry. These innovations have transformed technological innovation from a purely in-house activity in to a more collaborative one. Finally, transportation equipment corporations (especially car manufacturers, aircraft and aerospace firms, and mass transportation producers) are among the largest R&D spenders in all advanced countries. Thus, the potential for R&D economies of scale through technical collaboration is considerable for these firms.

TECHNICAL ALLIANCES IN TRANSPORTATION EQUIPMENT

In the transportation equipment industry, technical alliances tend to be one of two main types: they are either vertical or horizontal alliances. Vertical alliances include those of the transportation equipment producers with their suppliers of parts and components, with materials producers (aluminum, plastic, glass, and the like), and with their main clients (commercial airlines, defense firms, and other government departments and research laboratories). Horizontal alliances include those between independent transportation equipment manufacturers. The ten-year, DM2-billion German maglev, presently at the testing stage, is an example of technical association between three large firms (Thyssen, AEG, and MBB) that are conducting a large, expensive, and risky research project.[1] Electric car projects, particularly those researching new materials for advanced batteries, are conducted collectively in France by Peugeot, Renault, and Electricité de France, and in the United States by GM, Ford, and Chrysler, with the help of the U.S. Department of Energy (Stix 1992b). The national or international location of assemblers, suppliers, or clients is not usually a major consideration, except for military equipment, since the transportation equipment industry is largely multinational: international collaborative projects, are common in aircraft research and development, for example.

Vertical relationships between transportation equipment manufacturers and their first-tier suppliers often include frequent consultations and long-term projects between independent partners. Perfect compatibility of subsystems and quality control provide the main reasons for these vertical collaborations, but cost and risk factors are also at stake. Technological alliances in the commercial aircraft industry have been well studied and documented. The development by aluminum

---

1 Maglev (for "magnetic levitation") is a high-speed train running at a maximum of 450 km/h on a cushion of air (Stix 1992a).

corporations of high-strength alloys for aircraft producers is a striking example of technical collaboration between metal refiners and transportation equipment manufacturers (Pruitt and Smith 1986). Vertical alliances also include consortia and collaborations between engine and aircraft manufacturers for the development of the new motors: one example was the joint venture in the 1970s between General Electric of the United States and SNECMA of France for the development of a high-powered CFM56 engine (Mowery 1987). In these alliances, the technological thrust comes mainly from the suppliers, but they are responding to the demands of the customer. The introduction of electronic controls and telecommunication systems in transportation equipment is an example of this pattern.

Horizontal alliances include technical collaboration for the development of new electric automobiles, aircraft, and subway or railway cars, along with the corresponding parts, controls, and components. The R&D costs and sharing of risks are of vital importance here, since new transportation systems are time-consuming, expensive, and risky to develop and commercialize. The Airbus consortium of 1969, which brought together French (Aérospatiale), German (Deutsche Airbus), British (British Aerospace), Dutch (Fokker), Belgian (Belairbus), and Spanish (CASA) partners to develop a new family of aircraft is one of the most striking examples of horizontal collaboration in transportation equipment manufacturing (Claverie 1991). Other examples include Eurocopter (the European helicopter project) and the AEG-Westinghouse joint venture of 1988. The three large American car manufacturers have organized some ten consortia for research on paints, batteries, crash dummies, and the like (Coy 1993).

The Canadian Industry

The Canadian transportation equiment industry is no exception to these general rules. But the Canadian industry from those in other developed countries in the relative importance of the diverse segments of the industry.

The automobile industry is the largest segment in terms of sales. Canada produces nearly a million cars a year, but its R&D is lagging. Comparative figures give an idea of the missing expenditures. In 1990, car producers in the United States spent nearly Can$10 billion on R&D. In Canada, however, the whole motor vehicle industry spent less than Can$100 million (Statistics Canada 1991). Canadian car producers, who are all under foreign control, are technological borrowers from their parent companies in the United States, Japan, and Korea. The potential for R&D collaboration in the Canadian automobile industry is thus very limited, as in the demand for innovative new materials and electronic equipment. In the shipbuilding industry, Canada is also a technological laggard, performing little R&D.

However, Canada does have two major niches in which industrial R&D is relatively important: railway transportation equipment (including urban mass-transportation) and aircraft. In both cases, the industry is composed of a handful of large systems-assemblers, a few large first-tier suppliers, and hundreds of small and medium-sized parts and components manufacturers (mainly second- and third-tier suppliers). Vertical integration is unusual, and, as in the U.S. aircraft industry, the assemblers and suppliers are linked by a complex web of contractual relationships (Mowery and Rosenberg 1982; for Canada, see Horson 1983; Sullivan and Milberry 1989). There is entry at the suppliers' level, but it is almost nonexistent at the first-tier level and at the level of the large assemblers.

In 1989, nineteen aircraft and parts manufacturers in Canada performed R&D valued at Can$416 million. In the rest of the transportation equipment industry, consisting mainly of railway and subway manufacturing, eighteen corporations performed R&D valued at Can$114 million (Statistics Canada 1991). These thirty-seven corporations made up our largest potential population. The automobile manufacturers were excluded, since their current, total intramural expenditures on R&D were only 0.2 percent of sales. The structure of the Canadian industry forced us to choose a few large firms for our interviews and a sample of SMEs that were suppliers to them.

The federal government has always been a key actor in this industry, as a privileged buyer through public procurement, as a major supplier of funds for R&D, and as a regulator of technical standards and users' performance (DeBreson, Niosi, et al. 1991).

THE RESEARCH

Characteristics of the Firms

We interviewed 20 corporations involved in transportation equipment manufacturing across Canada. Our respondents included both large assemblers and small and medium-sized suppliers. They had been in business for 30 years, on average, but the largest had been operating, on average, for fifty years.

The Alliances

The corporations interviewed had formed an average of three technological alliances (Table 58), and 55 percent of the firms had made

Table 58
Number of transportation equipment alliances per firm

Number	Frequency	Percentage
1 to 4	11	55
5 to 9	4	20
10 or more	5	25
Total	20	100

Table 59
Number of partners in transportation equipment alliances

Number	Frequency	Percentage
2	11	55
3 to 9	4	20
10 or more	5	25
Total	20	100

Table 60
Managing the transportation equipment alliance

Type of management	Frequency	Percentage
Leading member	14	70
Collective decisions	6	30
Total	20	100

between one and four collaborative agreements. Development was the exclusive goal of almost all the alliances. In only one case was fundamental research (on advanced materials) the main objective.

A majority of the alliances (55 percent) had only two members, but some very large partnerships had many members (Table 59). There were no joint ventures in this industry; more flexible memorandums of agreement were used to formalize the collaboration.

Open-ended alliances without time limits were common (in 30 percent of the cases), as were short- and medium-term alliances lasting from one to three years (55 percent). Much longer commitments of four years and more were uncommon (15 percent).

Seventy percent of the alliances were managed by the leading members, who were usually large systems-assemblers or first-tier suppliers. The rest were managed through collective decision making by all members (Table 60).

Table 61
Solving the intellectual-property dilemma for transportation equipment

Solution	Frequency	Percentage
Common property	10	50
Leader is owner	8	40
Belongs to each partner	2	10
Total	20	100

[a] Multiple choice question.

Intellectual-Property Strategies

The intellectual property was shared by all partners in 50 percent of the cases, but it belonged exclusively to the leader in 40 percent. A distinctive pattern emerged for the division of intellectual property. Systems-integrators (large assemblers of transportation equipment) usually kept the intellectual property that resulted from their R&D collaboration with smaller subcontracting firms. But when firms of similar size (either large or small) worked together on a project, either the result belonged collectively to all firms or each partner kept the intellectual property based on its own research results (Table 61).

Patents had been used to protect common property in only three cases. In another two cases, the companies expected to obtain patents for the results of the collaboration. Generally, process technology was protected through secrecy and product technology through patents if the product improvement was considered patentable.

Domestic partnerships were the most frequent form, and they were vertical alliances of the assembler-supplier type. But there were some major horizontal, assembler-assembler alliances between Canadian and foreign assemblers, either American or European firms.

Partners were chosen for their technical capacity (in 90 percent of the cases), complementarity (45 percent), and previous knowledge (45 percent). As in the other industries, size was not mentioned as a criterion for chosing partners (Table 62).

Collective R&D received public sector financing very frequently: 75 percent of the alliances received some public funds; only 25 percent did not (Table 63).

Advantages and Difficulties

According to the firms, technical alliances were formed to reduce R&D costs (in 75 percent of the cases), to open windows on new technologies

Table 62
Explaining the choice of partners in transportation equipment alliances

Reason	Frequency	Percentage[a]
Technical capacity	18	90
Complementarity	9	45
Previous knowledge	9	45
Other reasons	7	35
Size	0	0

[a] Multiple choice question.

Table 63
Public financing in transportation equipment alliances

Financing	Frequency	Percentage
Private and public	15	75
Private only	5	25
Total	20	100

Table 64
Reasons for transportation equipment alliances

Reason	Frequency	Percentage[a]
Economies of R&D	15	75
Open windows on new technologies	12	60
Response to demands from users, suppliers	12	60
New markets	10	50
Complementarities	8	40
Other reasons	8	40

[a] Multiple choice question.

(60 percent), to respond to a demand from users or suppliers (60 percent), to look for new markets (50 percent), and to achieve complementarities (40 percent) (Table 64).

All our firms found advantages in the alliances they conducted. The complementary knowledge captured by the firm was the most important advantage (in 85 percent of cases), followed by the possibility of bringing new products to the market (50 percent). Neither, R&D diversification (20 percent) nor patents (10 percent) seemed very important: transportation equipment is an industry in which innovation is better protected by secrecy than by patents and in which large, fixed capital investments and related irreversibilities are frequent and diversification is low (Table 65).

Table 65
Advantages of transportation equipment alliances

Advantage	Frequency	Percentage[a]
Complementary knowledge	17	85
New products on the market	10	50
New customers	7	35
Accelerated innovation	6	30
Other advantages	6	30
Financing	5	25
R&D diversification	4	20
Patents	2	10

[a] Multiple choice question.

A majority of firms (80 percent) did not have any difficulty during the negotiation of the agreement, but most of them did have problems during its *implementation* (55 percent): communication problems and slow public sector funding decisions were the major difficulties. Transaction costs also exist in this industry, but they are not decisive factors.

Public financing did not generally create problems during the negotiation of the agreement, but it did during the implementation of the agreement. However, two of the four firms (two out of twenty, or 10 percent of the sample) that pointed to difficulties did identify slow negotiations with government as the main problem in settling the agreement. The other difficulties were not linked to government: they were problems with suitable partners (one case) and problems of confidentiality when university partners were involved.

How important were the results of the collaboration? Both in the short and long term, they were very important for all the firms we studied (Table 66). The explanation is to be found in the type of R&D the firms are doing in this industry: it is very different from some of the more basic research in the other technologies, because it is always development and thus always close to products for the market and close to customers.

Explanatory Variables

The size of the firm was closely related to its R&D behaviour. The average, unweighted expenditure on R&D by large firms (with 1000 employees and more) was valued at 3 percent of sales. For small and medium-sized firms, the figure was 10 percent (Table 67). Our figures for the industrial breakdown of expenditures correspond closely to

Table 66
Assessing the results of R&D in transportation equipment alliances

Assessment	Short term	Long term
Not important (0–5)	—	—
Important (6–7)	—	—
Very important (8–10)	20 (100%)	20 (100%)
Total	20 (100%)	20 (100%)

Table 67
R&D effort relative to transportation equipment sales

Size of firms	Frequency	R&D expenditures as a percentage of sales
Large	8	3
SMES	12	10
Total	20	7

Table 68
Cooperative R&D effort in transportation equipment relative to total R&D expenditures

Size of firms	Frequency	Cooperative R&D as a percentage of total expenditures
Large	8	20
SMES	12	24
Total	20	22

Statistics Canada's estimates of current intramural R&D expenditures as a percentage of performing company sales: in 1989, aircraft and parts manufacturers spent 14.8 percent of their sales on R&D, while all other transportation equipment manufacturers spent 5.2 percent (Statistics Canada 1991).

As in all the other sectors, larger firms devote a smaller percentage of their total R&D expenditures to collaborative R&D than do SMES. Cooperative R&D represented 20 percent of the total R&D expenditures in large firms (systems-assemblers) but an average of 24 percent in SMES (Table 68). Thus, again, there is strong evidence that alliances are a means of achieving R&D economies of scale. Larger firms, however, conducted more collaborative projects (an average of 20) than SMES (an average of 4).

There was scattered evidence of the adoption of new management methods, including increasing numbers of alliances between large assemblers and their regular suppliers. There was less evidence of the involvement of users (private transportation companies or public commissions) in those alliances. In fact, few users were found in the alliances, but of course interactive relationships do not necessarily take place in formal R&D collaborations like alliances.

Regional, National, and International Alliances

National collaborations were the most widespread in transportation equipment. Most often they involved private and, sometimes, public corporations working together in the development of new systems or components. These government-funded alliances usually included universities and public laboratories. The Alcatel Canada consortium (discussed below) is a good example of this type of collaboration.

International collaborations were most often formed by large systems-assemblers (producers of urban mass-transit equipment, railway and aircraft manufacturers). These were large and costly international alliances, with long-term horizons and little government funding. The alliance of Pratt and Whitney Canada with MTU (discussed below) is a case in point of this kind of collaboration.

Regional collaborations were equally divided between private company-to-company and industry-university agreements, with some public laboratory involvement (Tables 69 and 70). The Métro-Plus consortium (discussed below) is an interesting case of this type of agreement.

Industry Differences

The aircraft industry is composed of a few large assemblers (now reduced in numbers, because in 1991, Montreal's Canadair acquired Toronto's De Havilland) surrounded by several dozen first-tier suppliers. Vertical technical collaboration takes place between these systems-assemblers and their main suppliers of parts and components, many of which are not located in Canada. For instance, the Canadair Challenger 600 business jet, launched in 1978, is propelled by AVCO Lycoming turbines made in the United States, while the Challenger 601, launched in 1983, uses American-made General Electric turbines. On the other hand, De Havilland's recent Dash 8 uses Pratt and Whitney engines designed and built in Montreal. The cases are indicative of the pattern of technical alliances, because collaboration between aircraft assemblers and engine manufacturers is the most intense and vital in the whole industry (Canadair 1985; Sullivan and

Table 69
Transportation equipment alliances according to the geographic distribution of the
majority of partners

Scope	Frequency	Percentage
National	63	57
International	22	20
Regional	25	23
Total	110	100

Table 70
Membership of international and regional transportation equipment alliances

Partners	Scope	
	International	Regional
Firms	20	13
Universities[a]	2	12
Total	22	25

[a] Includes government laboratories.
Note: Fisher's Exact Test (2 tail) probability = 0.000.
   Phi = 0.90.

Milberry 1989). There is little collaboration among the suppliers
themselves, or between the first- and the second-tier suppliers: the
latter seldom perform any R&D. Large assemblers also collaborate with
universities and government laboratories on materials and fuels.

The production of railway and subway cars is similar: a few large
assemblers (also recently reduced in numbers when Quebec's mass-
transportation producer, Bombardier, acquired Ontario's subway manufacturer, UTDC) collaborate with their closest first-tier suppliers. In
this segment of the industry, however, national and regional collaborations are more widespread, since this more established technology
requires fewer specialized technological inputs. There are also international collaborations between large assemblers in this specific sector
of the industry.

CASE STUDIES OF CANADIAN ALLIANCES

Bombardier and the Métro-Plus Consortium

In 1990, the Quebec government announced that its Technology
Development Fund (TDF) would finance Métro-Plus, a consortium led

by Bombardier, the largest Canadian-owned transportation equipment manufacturer. A diversified enterprise, Bombardier includes Canadair, the world's largest manufacturer of business jets, and a mass transit division with 40 percent of the North American subway market. The Métro-Plus consortium is pursuing a three-year, $20-million R&D project to improve the reliability, price, safety, and performance of subway cars. The consortium was formed as a result of a $20-million order from the New York Transit Authority, received by Bombardier in 1988, to manufacture nine prototypes of high-technology subway cars. The consortium will develop microprocessor-based subsystems: for example, signal multiplexers and on-board test equipment. Bombardier participates in this cooperative project with two other private Montreal-area companies, Tech Rep Electronics and Pocatec, and two research organizations, CRIQ (the Centre de recherche industrielle du Québec, the public laboratory of the Quebec government) and the Centre spécialisé de technologie physique of La Pocatière College.

The TDF subsidy for Metro-Plus totals $5.7 million, or 28 percent of the total estimated project costs, and another 10 percent in credits. Bombardier and the partners are funding the balance. Bombardier will bring most of the private funding to the consortium and keep all the intellectual property stemming from the collaboration.

Alcatel, CN, and the ATCS Consortium

A railway-control technology consortium that includes Canadian National, Motorola Canada, and Vapor Canada of Montreal was formed in 1988 under the leadership of the SEL division of Alcatel Canada. Alcatel Canada, the Canadian subsidiary of the giant French telecommunications producer, Alcatel Alsthom, is the world's largest telecommunications equipment manufacturer and an active partner in the French high-speed train – TGV (*train à grande vitesse*). To further its objectives in Canada, Alcatel bought Canada Wire and Cable from Noranda in April 1991.

Canadian National Railways (CNR), Canada's largest railway company, was the first user involved in the consortium. A $20-million R&D project is under way to develop, install, and test an Advanced Train Control System (ATCS) on a CNR rail line. SEL is the systems integrator responsible for the computers both on board the locomotives and in the central office. Motorola engineers the radio communication network, and Vapor Canada develops the computer displays and keypads for the locomotive cab, together with the transponders. The ATCS would be the first of its kind in North America, and it would put the Canadian consortium in the lead over U.S. and Canadian competitors.

Industry, Science, and Technology Canada (ISTC) will finance up to 40 percent of the eligible costs of R&D, and CNR will finance the costs of installation and testing. The three main partners will fund the balance (*Research Money*, 22 February 1988).

Pratt and Whitney Canada and MTU

Pratt and Whitney Canada (P&WC), which is the subsidiary of the United Technologies Corporation of the United States, is the world leader in gas-turbine engines for small aircraft. Starting in the 1960s, P&WC developed a successful engine (the PT6) that is widely used in small aircraft across the world. In the early 1980s, P&WC started the design of a large turbofan, the PW300. By 1985, the basic dimensions of the PW300 design were settled; the development costs were estimated at Can$500 million; and P&WC was seeking a partner. Motoren and Turbinen Union Gmbh (MTU), the subsidiary of the German group Daimler-Benz, agreed to participate in the project and assume 25 percent of the research costs. MTU was chosen because of its expertise in aircraft-engine development: it also participates in two other lage strategic alliances with European partners, one to develop a fighter engine and the other a commercial turbofan. MTU became responsible for the design, development, and production of a PW300 low turbine. Under a loan repayable with future sales, the Canadian government financed 26 percent of the development costs of the PW300 (Sullivan and Milberry 1989). No details were published about the intellectual-property agreement linking P&WC and MTU.

The first two cases, Métro-Plus and ATCS, are good examples of national alliances: they are smaller, government-funded R&D projects. The link between PCW and MTU, with its larger budget, longer time frame, lower level of government support, and major emphasis on development, is typical of international alliances.

CONCLUSION

Transportation equipment manufacturing in Canada is in some ways similar to and in many ways different from the three generic technologies examined in previous chapters. Like the others, small and medium-sized transportation equipment companies undertake a larger R&D effort in relation to total sales and a stronger collaborative effort in relation to total intramural R&D expenditures than do large corporations. And the most expensive and most visible projects involve the largest corporations in the industry. In transportation equipment as in the other cases, companies forming alliances were successful in their search for complementary knowledge, new products, new technologies, and R&D economies.

The differences were, nevertheless, striking. Contrary to the pattern in the previous technologies, development in almost exclusively the goal of the alliances. Transportation equipment is, in a sense, a mature industry, very different from the previous industries and one that already has its established products and markets. Thus, alliances in transportation equipment are very important, both in the short and long term. They are also centred more on the development of products and processes and less on precompetitive research. In this industry there are also large, established companies able and willing to manage the collaborations and smaller firms acting as their long-term suppliers of parts and components. Unlike biotechnology, this is an old, well-structured industry with less technological turbulence and few direct scientific inputs. Nevertheless, the industry provides a demanding market for innovations in new materials and electronic controls and, as a result, it is involved in induced technical change.

7 Implications for Government and Business

Industrial research and development in Canada started early in the twentieth century, often on the basis of experience gained in the testing laboratories of a few chemical firms, for example. Over the years, a "linear" model of R&D emerged: the corporate research department produced new or improved processes and products and then sent them to manufacturing for analysis and eventual incorporation into production; the marketing and financial departments entered only later in the process of innovation. Conventional research was an in-house activity, basically conducted behind closed doors, with only occasional external technical collaboration. So the present collaborative revolution brings new models of industrial research and new management schemes. And it has important implications for governments.

IMPLICATIONS FOR GOVERNMENT

Government support for technical innovation in Canada started well before the founding of the National Research Council of Canada in 1916. One of the main goals was to provide a substitute for the weaker effort of Canadian industrialists and farmers through the creation of public laboratories. In-house research by the federal and provincial governments was conducted mainly for the benefit of small and medium-sized enterprises, like farms, that were unable to sustain their own research activities. Public laboratories were complemented with founding programs for industry and universities. Thus, from the start there was some kind of government-industry collaboration. But since the 1980s, government policies have been changing rapidly to incorporate cooperative R&D. In the early 1990s there were nearly 130 programs for cooperative research in Canada, but no comparative evaluation of these programs has ever been conducted. Our research is not primarily an evaluative exercise, but nevertheless, several major policy implications emerge from our study.

1 *Canadian corporations find important benefits in technical alliances.* One major conclusion of our study is that the trend towards collaborative research is well established, that it encompasses firms of all sizes, and that the private sector initiative is overwhelming. The increased reliance on technical alliances by high-technology firms and their own evaluations of the importance of technical cooperation show that this form of R&D management is beneficial and that it will probably become even more prevalent in the next few years. Government support for cooperative innovation is useful to and welcomed by the private sector, even if public funds are unevenly distributed in cooperative research. The increasing public support for technical cooperation is also probably a beneficial trend.

2 *Technical alliances are not simply government-promoted agreements.* A large but variable proportion of Canadian technical alliances, and probably some of the largest, are not funded by any government program. There are many economic considerations (R&D economies of scale, increasing risk and uncertainty in technological innovation, a more competitive national and international environment, growing spillovers and complementaries of technological development, accelerated innovation, product standards) motivating Canadian and foreign corporations to undertake cooperative R&D. Government incentives are probably facilitating the trend, but not creating it.

3 *Many government programs compete to subsidize collaborative R&D.* The number of federal and provincial programs developed in the last five years to subsidize technical collaborations has grown very rapidly, and many programs, research laboratories, and government institutions have been reorganized to finance, assist, and promote technical alliances. But there are indications that "one-stop shopping" is becoming a major policy issue in the federal government. Even though problems of duplication are evident at the provincial and federal levels (Gagnon 1990), the companies we interviewed were often unaware of many departments (both federal and provincial), agencies, and crown corporations offering assistance for cooperative R&D. Thus, some kind of coordination and simplification of structures may be necessary within each level of government and

between federal and provincial agencies. However, some degree of competition between government agencies may also be useful in order to ensure that important projects are supported. It was not the goal of our research, however, to determine the optimum level of government support and the optimum number of public agencies that Canadian consortia require to maximize the general welfare.

4 *Government-financed alliances are different from privately organized alliances.* Programs emanating from federal and provincial government departments are biased towards and, in the final analysis, more appropriate to the financing of large consortia conducting basic and applied research with a wide variety of industrial, university, and public members. The precompetitive bias of publicly funded research, the aversion to picking winners, diffusion considerations, and the sheer publicity value of large projects all contribute to the bias of government agencies towards larger consortia. Thus, government-backed alliances are usually formed through long and open negotiations between many organizations with very different goals and means. Intellectual-property sharing among government laboratories, universities, and private firms and also among very different industrial partners seems to be more difficult within this kind of consortium than in purely private alliances, because of several factors: the more rigid rules about intellectual property resulting from research conducted within government laboratories and universities, diverging goals, the conflicting priorities and different time frames of the research units involved, and the usually less precise parameters of precompetitive research. On the other hand, difficulties are less frequent in privately organized and financed partnerships that have little government support, because their objectives are more precise and the number of members is reduced: they are usually two-member teams whose goal is the development of new or improved products and processes. One may conclude that this is a normal state of affairs, since governments must support the more fundamental research and its industrial applications, while the more costly and less risky phases of development are left to private corporations.

5 *Technical alliances are not presently increasing economic concentration.* Of the alliances we studied, 10 percent, at the most, were joint ventures. Equity participation among partners was unusual, with the exception of the stakes held by large enterprises in small and medium-sized DBCs and a few advanced materials start-ups. When our information was collected, the overwhelming majority of the corporate partners were still independent, and we are not aware that the picture has changed significantly since then.

Nevertheless, barriers to entry could be erected in the future on the basis of present-day alliances. Collaborative rivals could merge and increase economic concentration. Or they could extend their technical cooperation to other areas and collude on such matters as price fixing or market sharing. Vertically linked allies could exclude would-be entrants from any specific sector of the market. Or more simply, the benefits of cooperative innovation by some firms may reduce the competitiveness of other firms de facto and force their exit from the market, thereby increasing industrial concentration (Mytelka and Delapierre 1987).

Policy Recommendations

Some policy recommendations can be drawn from these observations.

1 Governments should maintain their support for collective innovation, but they should a more balanced emphasis on the creation of smaller alliances, since the transaction costs associated with large consortia may be too burdensome and may cause inefficiencies in the use of funds and the composition of cooperative groups.
2 Government laboratories and universities should permit a variety of intellectual-property solutions. Since 1990, government laboratories have accepted this new rule, and public agencies are no longer the sole owners of all technology resulting from R&D conducted in their facilities. But industry-university collaboration is still hindered by the problem of appropriation. Several solutions (including collective ownership of all R&D, individual ownership of each company's R&D followed by cross-licensing, and "leader-takes-all") should be considered from the start in any negotiation between universities and industry. New routines must be implemented to ensure that government laboratories and universities are major – and not simply auxiliary – partners in Canadian alliances.
3 While corporations prefer to form alliances aimed at "development," academic researchers are interested in basic research. These diverging goals cannot always be accommodated, since the intellectual-property results have different values for different organizations: university researchers prefer publication, but firms prefer secrecy. Consequently, the inclusion of universities and government laboratories in the partnerships, though useful in many cases, should not be a condition of eligibility for most programs.
4 Because R&D is a corporate tool used to achieve competitive advantage, private companies prefer not to announce their innovative activities publicly, and agreements involving secrecy and confidentiality are often used to protect R&D results from free riders and

imitators. Thus, government support is more likely to be welcomed by business if it is less publicized.

5 Governments should conduct regular evaluations of their R&D programs, particularly newly implemented cooperative programs. The potential risks of collusion or merger among partners, the potentially high transaction costs of alliances, repeated intellectual-property litigation, government failures in cooperative project selection, and eventually diminishing support for the less fashionable – but highly important for Canada – resource-processing industries all call for a close scrutiny of public programs promoting collective innovation.

6 Because international alliances are costly, risky, and difficult to evaluate and because their benefits are even more difficult to appropriate within national frontiers, governments should give only restrained support to international partnerships, at least in the Canadian context, unless they are necessary and clearly conducive to the fulfillment of long-term national goals, as in the aerospace industry or in health, environmental, and alternative-energy research. Privately organized international technology developments should be funded by private firms.

IMPLICATIONS FOR BUSINESS

Technical collaboration may improve the efficiency of R&D, because corporations may reduce research costs by spreading expenditures across a larger number of partners and projects, by avoiding duplication, acquiring the complementary knowledge and facilities of partners at marginal cost, reducing risk and uncertainty, standardizing products outright, and designing products and processes directly from clients' specifications and demands.

Technical collaboration has costs and risks. As with any negotiation, there is the legal cost of the contract, since the cooperative partners need to protect themselves against eventual information leaks, cheating, and potential free riders. The negotiations must determine the contribution of each partner and the distribution of future results; partners must also agree on a management structure for the cooperative project. The more partners there are and the more divergent their goals and affiliations, the more costly and longer the negotiations will be. There are also direct costs of communication and transportation among the partners, both during the negotiations and in the course of implementating the project. The new collective governance structure also adds its own costs.

Corporations conducting collaborative research have adopted one of several management and intellectual-property sharing schemes.

The management schemes are most often democratic and collective: (*a*) they may range from simple surveillance committees for small collaborative projects to fully fledged boards of directors for large permanent consortia; (*b*) they may divide R&D among independent parties with periodic meetings of research managers to check the evolution of the different parts of the project; or (*c*) they may adopt a "leader-manages-all" solution in which a leading company provides most of the resources and has a major interest in supervising the project.

Similarly, the most frequent result-sharing schemes include either collective property rights to all information stemming from the collaboration, a division of labour and appropriation by each partner of its own R&D results (usually followed by cross-licensing agreements), or "manager-keeps-all" schemes. More complicated solutions include divisions of intellectual property in which some partners (usually universities or public laboratories) become owners of the fundamental aspects of the research for publishing purposes, while the commercial applications belong to the industrial partners. Also, in some cases, leading partners keep the right to use the main applications for which the project was originally designed, while similar partners keep the right to develop new applications outside the specific industry or sector that the collaborative project was designed for.

In our sample, patents were used only in a minority of cases, mainly in biotechnology, to protect results, either because precompetitive research was the goal of the collaboration and the technology was not patentable or because secrecy was the preferred method of appropriation. Few patents and copyrights were owned collectively by the partners to the alliances in our sample. Most often, intellectual property was registered under the name of one of the partners (the leader) or under the name of the consortium, or when possible, the technology was split into different packages, each belonging to a specific member of the alliance.

Business Strategies

We draw some important conclusions and lessons for business from our analysis:

1 It is necessary to start any collaboration with a thorough calculation of the transaction costs (legal, transportation, communication, and managerial costs) of the project. If not effectively estimated and controlled, these costs may cancel most or all the advantages and economies gained from cooperation.

2 It is crucial to balance the risks of information losses and free riding against the risks of failure resulting from an inappropriate attempt by a firm to execute a project without outside help or, more simply, from opportunistic behaviour of the partners. Adequate legal safeguards and prudent management choices (for example, designing a division of R&D among the partners and conducting separate parts of the project in each partner's facilities) can reduce this type of risk.

3 Management structures should be appropriate. Larger, lengthier, costlier projects and those focusing more on strategic development require more stable environments if they are to be conducted properly. An R&D joint venture controlled by the partners may be preferable in these cases. Smaller, precompetitive projects can be dealt with through more flexible forms of organization such as contractual agreements (MOUs). The relative size and the scope of the activities of the partners will determine whether one company should be leader, whether the partners should conduct their portion of the collective project in isolation, or whether there will be a free flow of ideas between the independent partners under the supervision of a committee or a board.

4 It is most important to conduct a proper evaluation of the technology contributed "in kind" by each partner and of the financial and other contributions of each partner. This evaluation may be difficult, for several reasons. First, the technology may be protected through secrecy to avoid the problem of free riders, but because intellectual property is in itself difficult to evaluate, it may not be possible to identify the precise contribution of a technology to the income of a corporation without in fact revealing the technology. Second, the advantages brought to the alliance by intangible assets may be difficult to compare with those of tangible assets like distribution networks and financial contributions. Finally, since the value of intellectual property is based upon its present contribution to profits, the value of a technology that is only partially developed (and thus producing no income for its owner) may be even more difficult to assess. Small and medium-sized innovative corporations may suffer from a serious disadvantage when they bring their partially developed technology to alliances with large firms, because it may be difficult to evaluate the precise technical contribution of the SME to the partnership. Large corporations typically supply manufacturing technology and R&D facilities, marketing networks, and cash, all of which are much easier to assess than the usually untested technology of small biotechnology or advanced materials firms.

5 It is also crucial to devise a clear distribution of the intellectual-property results. The collective ownership of intellectual property

Table 71
Joint patenting in Canada, 1980–1989[a]

Type of owners	Nationality of joint owners		
	Canadians	Canadian and foreign	Total
Individuals	484	19	503
Governments	19	1	20
Companies	14	0	14
Mixed	0	113	113
Total	517	133	650

[a] Total number of patents granted between 1980 and 1989 = 195,222.
Source: Canadian Patent Office, Patdat Database.

seems appropriate for the results of fundamental and precompetitive research, but individual ownership of specific results by each partner seems preferable for technology resulting from applied R&D. Collective ownership presents more problems when the results are patented (or protected by copyrights) and licensing to third parties is considered. In most European countries for instance (but not in the United States) all co-owners must agree on any future sale of a technology protected by copyright. Also, joint owners of patents and copyrights are considered by law to constitute a single party and must respond collectively to any litigation concerning the technology. So when development is the goal of the collaboration, a precise division of intellectual property, coupled with explicit allotments of markets and applications, is an absolute necessity. Alliances of this kind usually include marketing, financial, and manufacturing agreements about what the partner(s) will manufacture, what distribution channels will be used, and what financial contributions are expected from each partner for future development work.

6 The mode of appropriation (whether through patents, copyrights, secrecy, or some other mechanism) is another key issue that needs to be resolved from the start, because even when they occupy the same industrial niche, companies do not always follow the same intellectual-property strategies. Even strategies for scientific and technical publication vary considerably from one firm to the other, even within the same industry. The general pattern, however, is that a very small percentage of joint patents are held by Canadian firms (see Table 71). In fact joint appropriation is rare, and joint ownership of the results of cooperative R&D is negligible.

7 Except in small industry-university collaborations, negotiations should be conducted at a high level within the organization, usually by vice presidents with responsibility for R&D, intellectual property,

or technical development. The more general implication for business is that in strategic technical alliances, the risks, the costs, and potential gains are greater than in a simple licensing agreement or contract research. Consequently, the more explicit and complete agreements are the best, since they reduce uncertainties and risks and thus the chances of difficulties and litigation during the implementation of the agreement or at the conclusion, when the technology is already produced.

8 Conclusions

SUMMARY OF RESULTS

Since the early 1980s, Canadian companies of all sizes have been engaged in technical cooperation on a massive scale, among themselves and with foreign partners, and there is no indication that this trend will be reversed. Through networking, companies have sought specific advantages, such as economies of scale and scope in R&D. They have also hoped to capture complementary technology, design new and improved products, and obtain entry into new markets. More generally, the search for external information suggests that collective learning is a major goal of technological alliances.

Strategic partnerships seemed to be appropriate devices for attaining these goals, but the advantages sought from the alliances did not always correspond to those actually drawn. All firms, however, did find several advantages in their cooperative undertakings: new products and complementary knowledge were the most frequent.

Transaction Costs

We found in our study, as did Mariti and Smiley (1983) earlier, that avoiding transaction costs was not a major factor in the decision to engage in technical alliances: the benefits were expected to outweigh the costs of agreements, and, in fact, they did. But transaction costs did exist, and they do explain why economies of scale and cost reductions were not among the important advantages actually *obtained* from alliances, even though they were among the important advantages *sought* by the firms.

Patents

Patents were not generally a goal either sought by the firms in our study or in fact obtained, except in the biotechnology industry. Only large firms looked for patents in advanced materials, since only they could defend them against potential free riders. In other areas, companies preferred to protect their R&D though secrecy instead of patents. This result confirms Winter's (1989) previous finding that patents are an industry-specific method for protecting innovation.

Modes of Cooperation

Modes of cooperation varied from one industry to another. Table 72 summarizes the differences and similarities between industries. In biotechnology, only SMEs take the initiative to propose alliances to large firms who could develop and eventually use, manufacture, distribute, or test their inventions. Large transportation equipment assemblers organize vertical technical alliances with their smaller suppliers and horizontal ones with other large foreign assemblers. In electronics, the initiative comes from both large and small firms.

Geographic Scope

In every industry, alliances are mainly national in scope, but in electronics the percentage of foreign alliances is larger, reflecting Canadian competitiveness and internationalization in this area. Also, in electronics and transportation equipment, a few of these international alliances are financed by the Canadian government.

Goals of Alliances

In all sectors, partners are chosen for their technical competence. Development is the goal of most alliances, but not surprisingly, the percentage of alliances with development goals is much higher in transportation equipment, because it is a more established industry and a technological borrower from both advanced materials and electronics. Non-equity agreements are the predominant contractual form; they are much more frequent than joint ventures and probably reflect an intention to maintain a flexible alliance.

Table 72
Modes of cooperation in technological alliances

Variable	Electronics	Advanced materials	Biotechnology	Transport equipment
Initiative	Large firms and SMES	Both large firms and SMES	Only SMES	Large assemblers
Average number of partners	Three	Three	Two	Three
Scope (%)[a]	National (60)	National (91)	National (91)	National (80)
Base for partners choice	Technical capacity	Technical capacity	Technical capacity	Technical capacity
Dominant goal (%)	Development (49)	Development (78)	Development (68)	Development (95)
Predominant contract form (%)	Agreements (71)	Agreements (91)	Agreements (80)	Agreements (100)
Predominant management (%)	Leader (40)	Leader (55)	Independent R&D (61)	Leader (70)
Predominant intellectual-property form (%)	Collective (41)	Leader (42)	Collective (75)	Collective (50)
Main advantage sought (%)	New technologies (51)	R&D economies (58)	New markets (52)	R&D economies (75)
Main advantage drawn (%)	New products (63)	Complementary knowledge (58)	Complementary knowledge (58)	Complementary knowledge (58)
Main difficulty (%)	Intellectual property (25)	Intellectual property (28)	Intellectual property (12)	Government red tape (20)
Government funding percentage of projects	60	78	57	75

[a] Includes regional alliances.

Management Structures, Ownership of Results

In three of the four sectors, management by a leading firm is the most common form. Only in biotechnology, where university involvement is the highest and fundamental knowledge is very important, are alliances organized around highly independent R&D teams. Collective ownership of the results is more frequent in this sector than in the other three. Joint patents held by collaborative firms are rare, probably because of the legal problems mentioned earlier, such as joint responsibility for joint owners and restrictions on licensing to third parties.

Difficulties during Negotiations

Nearly half the firms experienced some kind of difficulty during the negotiations, but fewer had problems during the implementation of the agreement. In electronics, advanced materials, and biotechnology, the main problem was the sharing of the intellectual property. In biotechnology, where some alliances have very large R&D budgets, the second important problem was the financial contribution of the partners; however, this problem affected less than 10 percent of the firms. In the transportation equipment industry, a heavy user of government funds, red tape and longer negotiations over subsidies were more frequent than in other sectors.

Private and Public Funding

There was some government funding in every sector, but most of the funding was in transportation equipment. In all four samples, there was most frequently a mix of private and public funding. The percentage of projects supported entirely with private funding was highest in biotechnology and electronics and lower in transportation equipment and advanced materials. The size of the R&D budgets in these two industries is entirely different. Because innovative projects in transportation equipment are much larger and more time-consuming than in advanced materials, government allocations in transportation equipment are much more important, and subsidies require much more supervision from public agencies.

Size of the Firm and R&D Effort

In all four samples, the size of the firms was the main factor affecting collective R&D: the smaller the firm the larger was its R&D effort relative to its total sales and the larger was its expenditure on cooperative R&D

relative to its total expenditure on R&D. This finding is a major confirmation of the argument that alliances are organized in search of R&D economies of scale.

Size of the Firm and Technical Collaboration

The size of the firm also affected the number of collaborations: smaller firms entered into fewer alliances than larger ones, probably because of the reduced resources and capabilities of smaller firms and their less diversified technological portfolios. Also, specific subindustries behaved in different ways. In electronics, telecommunications equipment manufacturers were the most active in technical alliances, particularly in the international arena. Small manufacturers of semiconductors were also active with both national and international allies. In biotechnology, young DBCs in health care conducted more and larger alliances and were more active in transnational collaborations than other biotechnology firms. In advanced materials, large metallurgists and chemical firms were the most active in networking and the large metallurgists were more active in international cooperation. In transportation equipment, large assemblers in both the aircraft and the railway and subway sectors of the industry were doing collaborative research with their first-tier suppliers both in Canada and abroad. There were also some horizontal linkages among the first-tier suppliers within Canada and among the Canadian assemblers with foreign counterparts.

Reasons for Collaboration

Among the many suggested explanations for the formation of R&D alliances, the search for R&D economies of scale and complementary knowledge was basically confirmed by our research. Government funding also seems to be fuelling alliances in all sectors. Two explanations seem to apply only in specific industries: the search for common standards in electronics and the search for new methods of management in transportation equipment. Transaction costs do appear, but they also seem to be industry-specific and not important enough to cancel the advantages of alliances.

Regional, National, and International Alliances

Regional alliances usually included corporations and universities; national partnerships often involved government laboratories, universities, and crown corporations. Both regional and national alliances were government-funded. Their goals were mainly precompetitive. International alliances, on the other hand, seldom included universities (except in biotechnology), and they were intended, most often, to develop new or improved products and processes, and not to do fundamental or applied research. Since development was their goal, their time frames were longer, their budgets larger, and their organization more structured. International alliances included more cases of joint ventures and production and marketing agreements.

We conclude, more generally, that the new method of managing high-technology R&D is more than a passing fashion. It will probably be a long-term strategy for organizing innovation, and it will eventually win new industries as the new generic technologies are adopted by more traditional manufacturers. However, as Mowery and Rosenberg (1989) point out, cooperative research cannot be a substitute for in-house research but only a complement to it.

A NEW LOOK AT THEORY

R&D economies of scale were among the most important advantages sought from technical alliances. This finding gives some support to the arguments of industrial economics discussed in chapter 1. However, since economies of scale were sought not through mergers, as industrial economics would predict, but through specific arrangements – partnerships – that kept the enterprises independent, industrial concentration remained unaffected by technical partnerships. Moreover, static R&D economies of scale were more often mentioned by the partners as "advantages sought" than as "advantages obtained." Since nearly all companies have increased their collaborative activities, this gap suggests that, in the final analysis, other major benefits were realized by the cooperators.

Because institutional economics in the Coase-Williamson-Teece tradition excludes cooperation and recognizes only markets and hierarchies, it appears to be less useful for understanding these emerging cooperative organizational structures, which fall between markets and hierarchies. There is, in fact, some kind of consistency between the management structure adopted by the Canadian alliances and their expected transaction costs. Nevertheless, our study of alliances does show that some development of the institutional approach is necessary, since transaction-cost theory predicts that because they are more costly, these intermediate forms are less stable than either market transactions or hierarchies. Transaction-cost theory may, in fact, be less relevant than is usually assumed by institutionalists. Not only are they unable to explain why these new organizational forms developed in the first place; they are also unable to explain the exponential increase in the number of strategic alliances during the 1980s.

No one has tried to analyze technical alliances using an explicitly evolutionary approach, but we believe that evolutionary economics together with management theory can provide the basis for a deeper understanding of technical collaboration. Evolutionary economics argues that under appropriate external and internal conditions (for example, in the more turbulent economic environment resulting from the triple generic-technology revolution), organizational innovations such as alliances may emerge to ensure that firms are adapted to their new economic and political setting. These new forms may become permanent additions to the organizational routines of the firm. In sense, the predictions of evolutionary economics are more definite than those of transaction-cost theory. Since it depicts routines as additions to the organizational repertoire of the firm, evolutionism predicts that the present wave of technical collaboration will become a new and permanent trait of the organizational repertoire. In this sense, evolutionism is in direct opposition to transaction-cost theory, for which alliances are transitory forms that must eventually develop into hierarchies if the partners do not return to the market. It would be difficult at this stage to predict that alliances will evolve into more concentrated world oligopolies or even that they will unduly diminish the number of technical solutions available to industry.

The new techno-economic paradigm, which is closely linked to evolutionism, also sees alliances as linked to a major new phase in the development of technology and society: the present technological turbulence results from the information-technology revolution. Like the previous techno-economic paradigm, this one will set the stage for another one in which alliances and technical cooperation may not be central. Just as Taylorism and Fordism are now passing away, so too, eventually, will flexible, "just-in-time" production systems and technical alliances. In a sense, our results confirm the techno-economic paradigm: alliances are linked more to emerging, generic technologies than to traditional, established industries. But it would be hazardous to predict on the basis of our research that the intensity of the present technological revolution will diminish, or that technological collaborations will disappear.

The focus on regional networks of innovators contributes less to the explanation of alliances, since most technical collaborations are, quite simply, not regional organizations. The most important alliances, measured in terms of their funding, scope, or time frame, are either national or international organizations. Moreover, local alliances are quite often industry-university partnerships, and they generally focus on fundamental research and discovery. As such they should be considered more as *networks of inventors* than *networks of innovators*. It may be that regional cooperation is mostly informal and short-term (DeBresson and Amesse 1991), while national and international research is centred on more formalized collaborations that are supported by written long-term agreements. Nevertheless, regional economics and geography do correctly underline the externalities in technological alliances, but they incorrectly assume that these externalities need geographic proximity.

In summary, then, our study of technological (and non-technological) alliances draws elements from several different theories in an attempt to build an integrated framework based on evolutionary economics in the Nelson-Winter tradition. Alliances are seen as a key component of national systems of innovation; the R&D activities of economic units thrive within the specific institutional arrangements of the alliances. National systems of innovation are, like firms, complex, large-scale units of learning: within these national systems there is room for both competition and cooperation. Economics has traditionally underlined the advantages of competition. But alliances provide an important argument for the opposite side: economic efficiencies may result from cooperation among producers.

Bibliography

Aldrich, Howard E. 1979. *Organizations and Environments*. Englewood Cliffs, NJ: Prentice-Hall.

Allen, Robert C. 1983. "Collective Invention." *Journal of Economic Behaviour and Organization* 4(1):1–24.

Anchordoguy, Marie. 1989. *Computers Ltd. Japan's Challenge to IBM*. Boston: Harvard Business School Press.

Annett, William. 1991. "Macblo: Leaving Home." *Canadian Business*, July, 44–7.

Arrow, Kenneth. 1962. "Economic Welfare and the Allocation of Resources for Invention." In *The Rate and Direction of Inventive Activity: Economic and Social Factors*, edited by R.R. Nelson, pp. 609–25. Princeton: Princeton University Press.

Baumol, William J., and Y. Braunstein. 1977. "Empirical Study of Scale Economies and Production Complementarity. The Case of Journal Publication." *Journal of Political Economy* 85(5):1037–48.

Blackford, Mansel G. 1988. *The Rise of Modern Business in Great Britain, the United States, and Japan*. Chapel Hill: University of North Carolina Press.

Blair, John. 1982. *Economic Concentration*. New York: Harcourt, Brace and Jovanovitch.

Bleeke, Joel, and D. Ernst, eds. 1993. *Collaborate to Compete*. New York: Wiley.

Bonin, Bernard, and C. Desranleau. 1988. *Innovation industrielle et analyse économique*. Montreal: G. Morin.

Bothwell, Robert. 1988. *Nucleus. The History of Atomic Energy of Canada Ltd.* Toronto: University of Toronto Press.

Braillard, Philippe, and A. Demant. 1991. *EUREKA et l'Europe technologique*. Paris and Brussels: LGDJ and Bruylant.

Branscomb, Louis M. 1991. *America's Emerging Technology Policy.* CSIA Discussion Paper, no. 91–12. Cambridge: Kennedy School of Government, Harvard University.

– 1992. *S&T Information Policy in the Context of a Diffusion-Oriented National Technology Policy.* CSIA Discussion Paper, no. 92–12. Cambridge: Kennedy School of Government, Harvard University.

Breheny, Michael, ed. 1988. *Defense Expenditure and Regional Development.* London and New York: Mansell.

Campbell, Duncan C. 1990. *Global Mission. The Story of Alcan.* 3 vols. Montreal: Alcan.

Canada. Department of Industry, Science, and Technology. 1991. *National Biotechnology Business Strategy.* Ottawa: Supply and Services Canada.

Canada. Ministry of State for Science and Technology. 1980. *Biotechnology in Canada.* Ottawa: Supply and Services Canada.

Canada. National Research Council (NRC). 1985. *A Practical Perspective.* Ottawa: Supply and Services Canada.

– 1990. *The Competitive Edge. Long-Range Plan 1990–1995.* Ottawa: Supply and Services Canada.

Canadair. 1985. *Les 40 premières années. Une rétrospective des réalisations de Canadair, 1944–1984.* Montreal: Canadair.

Canadian Biotechnology Association. 1991. *Canadian Biotechnology Directory.* Ottawa: Canadian Biotechnology Association.

Chesnais, François. 1988. "Les accords de coopération technique entre firmes indépendantes." *STI Revue,* no. 4:55–132.

Claverie, Bruno. 1991. *La gestion des consortiums européens.* Paris: Presses Universitaires de France.

Coase, Ronald. 1937. "The Nature of the Firm." *Economica* 4(16):386–405.

Coghlan, Andy. 1993. "Engineering the Therapies of Tomorrow." *New Scientist* 138(1870):26–31.

Cohendet, Patrick, Jean-A. Héraud, and E. Zuscovitch. 1992. "Apprentissage technologique, réseaux économiques, et appropriabilité des innovations." In *Technologie et richesse des nations,* edited by D. Foray and C. Freeman, pp. 63–78. Paris: Economica.

Coombs, Rod, P. Saviotti, and V. Walsh. 1987. *Economics and Technological Change.* Totowa, NJ: Rowman and Littlefield.

Coy, Peter. 1993. "Two Cheers for Corporate Collaboration." *Business Week,* 3 May 1993, 34.

Dasgupta, Partha, and P. Stoneman. 1987. *Economic Policy and Technological Performance.* Cambridge: Cambridge University Press.

DeBresson, Christian. 1987. "The Evolutionary Paradigm and the Economics of Technical Change." *Journal of Economic Issues* 21(2):751–62.

DeBresson, Christian, and F. Amesse. 1991. "Networks of Innovators: A Review and Introduction to the Issue." *Research Policy* 20(5):363–79.

DeBresson, Christian, J. Niosi, and R. Dalpé. 1991. "Liaisons technologiques et contrôle étranger dans l'industrie aéronautique canadienne." In *Investissement étranger, technologie, et croissance économique,* edited by D. McFetridge, pp. 385–437. Calgary: University of Calgary Press.

Delapierre, Michel. 1991. "Les accords inter-entreprises: partage ou partenariat? Les stratégies des groupes européens du traitement de l'information." *Revue d'économie industrielle,* no. 55:135–61.

Diebold, John. 1990. "The Fibrer Optics Breakthrough". Chapter 7 in *The Innovators.* New York: Truman Alley.

Doern, G. Bruce, and R.W. Morrison. 1980. *Canadian Nuclear Policies.* Montreal: The Institute for Research on Public Policy.

Dosi, Giovanni, Christopher Freeman, R. Nelson, G. Silverberg, and L. Soete, eds. 1988. *Technical Change and Economic Theory.* London: Pinter.

Doz, Yves. 1988. "Technology Partnerships Between Larger and Smaller Firms: Some Critical Issues." In *Cooperative Strategies in International Business,* edited by Frank J. Contractor and P. Lorange, pp. 317–38. Lexington, MA: Lexington Books.

Dussauge, Pierre, B. Garrette, and B. Ramanantsoa. 1988. "Stratégies relationnelles et stratégies d'alliances technologiques." *Revue française de gestion,* no. 68:7–19.

Dutton, John M., and A. Thomas. 1985. "Relating Technological Change and Learning by Doing." In *Research on Technological Innovation, Management, and Policy,* edited by R.S. Rosenbloom, pp. 184–224. Greenwich, CT: Jay Press.

Eggleston, Wilfrid. 1978. *National Research in Canada. The NRC, 1916–1966.* Toronto: Clarke Irwin.

Elster, Jon. 1983. *Explaining Technical Change.* Cambridge: Cambridge University Press.

Ernst and Young. 1992. *Canadian Biotech/92.* Toronto: Ernst and Young.

Executive Office of the President. 1990. *The U.S. Technology Policy.* Washington, DC: GPO.

Forester, Tom. 1988a. *High-Tech Society. The Story of the Information Technology Revolution.* Cambridge, MA: MIT Press.

– 1988b. *The Materials Revolution.* Cambridge, MA: MIT Press.

Forrest, Janet E., and M.J.C. Martin. 1992. "Strategic Alliances Between Large and Small Research-Intensive Organizations: Experiences in the Biotechnology Industry." *R&D Management* 22(1):41–53.

Freeman, Christopher. 1987. *Technology Policy and Economic Performance. Lessons from Japan.* London: Pinter.

– 1988. "Japan: A New National System of Innovation?" In *Technical Change and Economic Theory,* edited by G. Dosi, C. Freeman, R. Nelson, G. Silverberg, and L. Soete, pp. 330–48. London: Pinter.

Freeman, Christopher, and C. Perez. 1988. "Structural Crisis of Adjustment: Business Cycles and Investment Behaviour." In *Technical Change and Economic*

Theory, edited by G. Dosi, C. Freeman, R. Nelson, G. Silverberg, and L. Soete, pp. 38–66. London: Pinter.

Freeman, John. 1990. "Ecological Analysis of Semiconductor Firm Mortality." In *Organizational Ecology*, edited by J.V. Singh, pp. 53–78. Newbury Park, CA: Sage.

Fusfeld, Herbert I. 1987. *The Technical Enterprise*. Cambridge, MA: Ballinger.

Fusfeld, Herbert I., and C. Haklisch. 1985. "Cooperative R&D for Competitors." *Harvard Business Review* 63(6):60–76.

– 1987. "Collaborative Industrial Research in the U.S." *Technovation* 5(4):305–16.

Gagnon, Pauline. 1990. *L'organisation de la politique scientifique du gouvernement du Québec et du gouvernement fédéral*. Quebec: Conseil de la science et de la technologie du Québec.

Gomes-Casseres, Benjamin. 1992. *International Trade, Competition, and Alliances in the Computer Industry*. Working paper, no. 92–044. Cambridge, MA: Harvard Business School.

Green, Ken. 1992. "Creating Demand for Biotechnology: Shaping Technology and Markets." In *Technological Change and Labour Markets*, edited by R. Coombs, P. Saviotti, and V. Walsh, pp. 164–84. London: Academic Press.

Gregory, Gene. 1987. "New Materials Technology in Japan." *International Journal of Materials Research and Product Technology* 2(1):42–51.

Gugler, Philippe. 1992. "Building Transnational Alliances to Create Competitive Advantage." *Long-Range Planning* 25(1):90–9.

Hagedoorn, John. 1990. "Organizational Modes of Inter-Firm Cooperation and Technology Transfer." *Technovation* 1(1):17–30.

Hagedoorn, John, and J. Schakenraad. 1991. "Inter-firm Partnerships for Generic Technologies: The Case of New Materials." *Technovation* 11(7):429–44.

Haklisch, Carmela. 1987. *Technical Alliances in the Semiconductor Industry*. New York: Center for Science and Technology Policy, Rensselaer Polytechnic Institute.

Hannan, Michael T., and J. Freeman. 1977. "The Population Ecology of Organizations." *American Journal of Sociology*. 82(5):929–64.

Hawthorne, Edward P. 1978. *The Management of Technology*. London: McGraw-Hill.

Hayek, Friedrich. 1978. "Competition as a Discovery Process." In *New Studies in Philosophy, Politics, Economics, and the History of Ideas*. Chicago: University of Chicago Press.

Hondros, E.D. 1986. "Materials. Year 2000." *International Journal of Materials and Product Technology* 1(1):54–77.

Horson, F. 1983. *The DeHavilland Canada Story*. Toronto: Canav.

Hounshell, David A., and J.K. Smith. 1988. *Science and Corporate Strategy. Du Pont 1902–1980*. Cambridge: Cambridge University Press.

Hull, James P., and P.C. Enros. 1988. "Demythologizing Canadian Science and Technology: The History." *Canadian Issues* 10(3):1–22.

Hybels, R.C. 1990. Occupational Politics in the Determination of Business Strategy: The Case of Strategic Alliances in the U.S. biotechnology Industry. Paper presented at the 10th International Conference of the Strategic Management Society, Stockholm.

Hydro Québec. 1991. *TAI Information* 7(1–2):7.

Industrial Biotechnology Association of Canada. 1991. *Canadian Biotechnology Directory, 1990–1*. Ottawa: Industrial Biotechnology Association.

Jéquier, Nicolas. 1974. "Computers." In *Big Business and the State: Changing Relations in Western Europe*, edited by R. Vernon, pp. 195–254. Cambridge: Harvard University Press.

Johnson, Chalmers. 1980. *MITI and the Japanese Miracle. The Growth of Industrial Policy, 1925–1975*. Tokyo: Tuttle.

Jorde, Thomas M., and D.M. Teece. 1989. "Competition and Cooperation: Striking the Right Balance." *California Management Review* 31(3):25–37.

Kay, Neil M. 1992. "Markets, False Hierarchies and the Evolution of the Modern Corporation." *Journal of Economic Behaviour and Organization* 17(3):315–33.

Killing, Peter J. 1983. *Strategies for Joint-Ventures Success*. New York: Praeger.

Kreiner, Kristian, and M. Schultz. 1990. Crossing the Institutional Divide. Networking in Biotechnology. Paper presented at 10th International Conference of the Strategic Management Society, Stockholm.

Langlois, Richard N., T.A. Pugel, C.S. Haklisch, R. Nelson, and W.G. Egelhoff. 1988. *Micro-Electronics. An Industry in Transition*. Boston: Unwin Hyman.

Larédo, Philippe, and M. Callon. 1990. *L'impact des programmes communautaires sur le tissu scientifique et technique français*. Paris: La documentation française.

Le Roy, Donald J., and P. Dufour. 1983. *Partenaires pour la stratégie industrielle. Le rôle des organismes provinciaux de recherche*. Ottawa: Science Council of Canada.

Lermer, George. 1987. *Atomic Energy of Canada Limited*. Ottawa: Supply and Services Canada.

Levy, Jonah, and R. Samuels. 1991. "Institutions and Innovation: Research Collaboration as Technology Strategy in Japan." In *Strategic Partnerships and the World Economy*, edited by L. Mytelka, pp. 120–48. Rutherford NJ: Fairleigh Dickinson University Press.

Lundvall, Bengt-A. 1988. "From User-Producer Interaction to the National System of Innovation." In *Technical Change and Economic Theory*, edited by G. Dosi, C. Freeman, R. Nelson, G. Silverberg, and L. Soete, pp. 349–69. London: Pinter.

– 1990. Explaining Interfirm Cooperation and Innovation–Limits of the Transaction-Cost Approach. Paper presented to the Workshop on the Socio-economics of Interfirm Cooperation, Wissenschafts Zentrum, Berlin.

Lundvall, Bengt-A. ed. 1992. *National Systems of Innovation.* London: Pinter.

Majundar, B.A. 1988. "Industrial Policy in Action: The Case of the Microelectronics Industry in Japan." *Columbia Journal of World Business* 23(3):25–33.

Mansfield, Edwin. 1977. *The Production and Application of New Industrial Technology.* New York: Norton.

Mariti, P., and R.H. Smiley. 1983. "Cooperative Agreements and the Organization of Industry." *The Journal of Industrial Economics* 31(4):437–51.

Markusen, Ann, P. Hall, and A. Glasmeier. 1986. *High-Tech America.* Boston: Allen and Unwin.

Marshall, Alfred. [1890] 1986. *Principles of Economics.* 8th ed. London: Macmillan.

McKelvey, Bill, and H. Aldrich. 1983. "Populations, Natural Selection and Applied Organizational Science." *Administrative Science Quarterly* 28(1):101–28.

McNeil, Ian, ed. 1990. *Transport.* Pt. 3 of *An Encyclopaedia of the History of Technology.* London and New York: Routledge.

Microelectronics and Computer Technology Corporation (MCC). 1991. *Fact Sheet.* Austin, TX: MCC.

Mowery, David. 1987. *Alliance Politics and Economics. Multinational Joint Ventures in Commercial Aircraft.* Cambridge, MA: Ballinger.

– 1989. "Collaborative Ventures Between U.S. and Foreign Manufacturing Firms." *Research Policy* 18(1):19–32.

Mowery, David, and N. Rosenberg. 1982. "The Commercial Aircraft Industry, 1925–1975." In *Inside the Black Box. Technology and Economics,* edited by N. Rosenberg, pp. 169–77. Cambridge: Cambridge University Press.

– 1989. *Technology and the Pursuit of Economic Growth.* Cambridge: Cambridge University Press.

Mytelka, Lynn. 1990. *Technological and Economic Benefits of ESPRIT.* Ottawa: Supply and Services Canada.

Mytelka, Lynn, and M. Delapierre. 1987. "The Alliance Strategy of European Firms and the Role of ESPRIT." *Journal of Common Market Studies* 26(2):231–54.

Mytelka, Lynn, ed. 1992. *Strategic Partnerships and the World Economy.* Rutherford, NJ: Fairleigh Dickinson University Press.

Nelson, Richard R. 1988. "Institutions Supporting Technical Change in the United States." In *Technical Change and Economic Theory,* edited by G. Dosi, C. Freeman, R. Nelson, G. Silverberg and L. Soete, pp. 312–29. London: Pinter.

National Advisory Biotechnology Committee. 1991. *Biotechnology Business Strategy.* 5th Annual Report. Ottawa: Industry, Science, and Technology Canada.

Nelson, Richard R., and S. Winter. 1982. *An Evolutionary Theory of Economic Change.* Cambridge, MA: Harvard University Press, Balknap Press.

Niosi, Jorge. 1993. "Strategic Partnerships in Canadian Advanced Materials." *R&D Management* 23(1):17–27.

Niosi, Jorge, ed. 1991. *Technology and National Competitiveness.* Montreal and Kingston: McGill-Queen's University Press.

Niosi, Jorge, B. Bellon, P. Saviotti, and M. Crow. 1993. "National Innovation Systems: In Search of a Workable Concept." *Technology in Society* 15(2):207–27.

Niosi, Jorge, and M. Bergeron. 1992. "Technical Alliances in the Canadian Electronics Industry." *Technovation* 12(5):309–22.

Niosi, Jorge, and R. Landry. 1993. "Les gouvernements et les alliances technologiques." *Gestion* 18(3):32–8.

Niosi, Jorge, and B. Bellon. 1994. "The Global Interdependence of National Systems of Innovation. Evidence, Limits and Implications." *Technology in Society* 16(2):173–97.

Norris, Keith, and J. Vaizey. 1973. *The Economics of Research and Technology.* London: Allen and Unwin.

Office of Technology Assessment (OTA), Congress of the United States. 1984. *Commercial Biotechnology.* Washington, DC: Government Printing Office.

Okimoto, Daniel. 1989. *Between MITI and the Market. Japanese Industrial Policy for High Technology.* Stanford: Stanford University Press.

Okimoto, Daniel, T. Susano, and F.B. Weinstein, eds. 1984. *Competitive Edge. The Semiconductor Industry in the U.S. and Japan.* Stanford: Stanford University Press.

Oshima, Keichi. 1984. "Technological Innovation and Industrial Research in Japan." *Research Policy* 13(4):285–301.

Peck, Merton. 1986. "Joint R&D: The Case of Microelectronics and Computers Technology Corporation." *Research Policy* 15(5):219–31.

Perlmutter, Howard V., and D.A. Heenan. 1986. "Cooperate to Compete Globally." *Harvard Business Review* 64(2):136–42.

Peters, Lois, P. Groenewegen, and N. Fiebelkorn. 1993. Public-Private Cooperation: A Comparison of European Support of Materials Technology and Biotechnology. Paper presented at the ASEAT Conference, April, University of Manchester.

Piore, Michael, and C. Sabel. 1984. *The Second Industrial Divide.* New York: Basic Books.

Pisano, Gary. 1990. "The R&D Boundaries of the Firm: An Empirical Analysis." *Administrative Science Quarterly* 35(1):153–76.

Pisano, Gary, and D.J. Teece. 1989. "Collaborative Arrangements and Global Technology Strategy: Some Evidence from the Telecommunications Equipment Industry." In *Research on Technological Innovation, Management, and Policy,* edited by R. Rosenbloom, pp. 227–56. Greenwich, CT: Jai Press.

Porter, Michael, 1980. *Competitive Strategy.* New York: Macmillan.

– 1985. *Competitive Advantage.* New York: Macmillan.

Province of Ontario. Ontario Premier's Report. 1988. *Competing in the Global Economy.* 3 vols. Toronto: Office of the Premier.

Pruitt, Bettye H., and G.H. Smith. 1986. "The Corporate Management of Innovation: Alcoa Research, Aircraft Alloys, and the Problem of Stress-Corrosion Cracking." In *Research on Technological Innovation, Management, and Policy*, edited by R. Rosenbloom, pp. 33–81. Greenwich, CT: Jai Press.

Queisser, Hans. 1988. *The Conquest of the Microchip. Science and Business in the Silicon Age*. Cambridge: Harvard University Press.

Rhea, John. 1991. "New Directions for Industrial R&D Consortia." *Research Technology Management* 34(5):16–26.

Rosenberg, Nathan. 1982. *Inside the Black Box: Technology and Economics*. Cambridge: Cambridge University Press.

Rosenberg, Nathan, and C. Frischtak, eds. 1985. *International Technology Transfer*. New York: Praeger.

Sapienza, Alice M. 1989. "R&D Collaboration as a Global Competitive Tactic. Biotechnology and the Ethical Pharmaceutical Industry." *R&D Management* 19(4):285–95.

Saviotti, Paolo, and J.S. Metcalfe, eds. 1992. *Evolutionary Theories of Economic and Technical Change*. Reading: Harwood.

Saxenian, Anna Lee. 1991. "The Origins and Dynamics of Production Networks in Silicon Valley." *Research Policy* 20(5):423–38.

Scherer, Frederic M. 1970. *Industrial Market Structure and Economic Performance*. Chicago: Rand McNally.

Schumpeter, Joseph. [1911] 1934. *The Theory of Economic Development*. Cambridge: Harvard University Press.

Schumpeter, Joseph. 1942. *Capitalism, Socialism, and Democracy*. London: Allen and Unwin.

Senker, Jacqueline, and W. Faulkner. 1992. "Industrial Use of Public Sector Research in Advanced Technologies: A Comparison of Biotechnologies and Ceramics." *R&D Management* 22(1):157–75.

Sercovich, Francisco, and M. Leopold. 1991. *Developing Countries and the New Biotechnology. Market Entry and Industrial Policy*. Ottawa: IDRC.

Sharp, Margaret. 1989. *Collaboration in the Pharmaceutical Industry*. DRC Discussion Paper, no. 71. Sussex: SPRU, Sussex University.

Siggel, Eckhart. 1987. "Learning by Consulting: A Model of Technology Transfer by Consulting Engineering Firms." *Canadian Journal of Development Studies* 6(1):27–44.

Simon, Herbert I., M. Egidi, and R. Marris, eds. 1992. *Economics, Bounded Rationality, and the Cognitive Revolution*. London: Elgar.

Smith, D.M. 1981. *Industrial Location. An Economic Analysis*. New York and Toronto: Wiley.

Soete, Luc. 1991. The Internationalization of Science and Technology Policy: How Do Nations Cope?. Working Paper, MERIT, Maastricht.

Statistics Canada. 1989. *Manufacturing Industries of Canada*. No. 31–203. Ottawa: Supply and Services Canada.

- 1991. *Industrial Research and Development Statistics 1989*. No. 88–202. Ottawa: Supply and Services Canada.

Steed, Guy, and S. Tiffin. 1986. *Une consultation nationale sur les technologies émergentes*. Ottawa: Science Council of Canada.

Stent, Gunther. 1993. "DNA's Stroke of Genius." *New Scientist* 138(1870):21–5.

Stiglitz, Joseph. 1987. "Learning to Learn: Localized Learning and Technological Progress." In *Economic Policy and Technological Performance*, edited by P. Dasgupta and P. Stoneman, pp. 125–53. Cambridge: Cambridge University Press.

Stix, Gary. 1992a. "Riding on Air." *Scientific American*, February, 104–5.

- 1992b. "Electric Car Pool." *Scientific American*, May, 126–7.

- 1993. "Concrete Solutions." *Scientific American*, April, 102–12.

Storper, Michael, and B. Harrison. 1991. "Flexibility, Hierarchy and Regional Development." *Research Policy* 20(5):407–22.

Sullivan, Kenneth H., and L. Milberry. 1989. *Power. The Pratt and Whitney Canada Story*. Toronto: Canav.

Tarasofsky, Abraham. 1984. *The Subsidization of Innovation Projects by the Government of Canada*. Ottawa: Supply and Services.

Teece, David. 1988. "Technological Change and the Nature of the Firm." In *Technical Change and Economic Theory*, edited by G. Dosi, C. Freeman, R. Nelson, G. Silverberg, and L. Soete, pp. 256–81. London: Pinter.

- 1989. Competition and Cooperation in Technology Strategy. International Business Working Paper IB-11, School of Business, University of California at Berkeley.

Teitelman, Robert. 1989. *Gene Dreams. Wall Street, Academia, and the Rise of Biotechnology*. New York: Basic Books.

Teubal, Morris. 1987. *Innovation Performance, Learning, and Technology Policy*. Madison, WT: University of Wisconsin Press.

Von Hippel, Eric. 1977. "The Dominant Role of the User in Semiconductor and Electronic Subassembly Process Innovation." *IEEE Transactions on Engineering Management* 24(2).

- 1987. "Cooperation Between Rivals: Informal Know-How Trading." *Research Policy* 16(6):291–302.

Walsh, Vivien. 1991. Demand, Public Markets, and Innovation in Biotechnology. Paper presented to the Symposium on Demand, Public Markets and Innovation, Montreal: CREDIT.

Watkins, Todd A. 1991. "A Technological Communications Costs Model of R&D Consortia as Public Policy." *Research Policy* 20:87–107.

Whitcomb, David. 1972. *Externalities and Welfare*. New York and London: Columbia University Press.

White, Lawrence J. 1985. "Clearing the Legal Path to Cooperative Research." *Technology Review* 87(5):39–44.

Williamson, Oliver. 1975. *Market and Hierarchies*. London: Free Press.

- 1985. *The Economic Institutions of Capitalism*. New York: Macmillan.

Willinger, Marc. 1989. "La diffusion des matériaux composites dans les systèmes complexes et l'intensification des relations interindustrielles." *Revue d'économie industrielle* 49:51–66.

Willinger, Marc, and E. Zuscovitch. 1988. "Towards the Economics of Information-Intensive Production Systems: The Case of Advanced Materials." In *Technical Change and Economic Theory*, edited by G. Dosi, C. Freeman, R. Nelson, G. Silverberg, and L. Soete, pp. 239–55. London: Pinter.

Wilson, John. 1991. The Future of Material Science: The Engineering Connexion. An Address to the IMI/NRC 75th Anniversary Symposium, Boucherville, Quebec.

Winter, Sidney. 1989. "Patents in Complex Contexts: Incentives and Effectiveness." In *Owning Scientific and Technical Information*, edited by V.W. Weil and J. Snapper, pp. 41–60. New Brunswick, NJ and London: Rutgers University Press.

– 1990. "Survival, Selection, and Inheritance in Evolutionary Theories of Organization." In *Organizational Ecology*, edited by J.V. Singh, pp. 269–297. Newbury Park, CA: Sage.

Woods, Stanley. 1987. *Western Europe, Technology, and the Future*. London: Croom Helm.

Yoshida, Kosaku. 1992. "New Economic Principles in America – Competition and Cooperation." *Columbia Journal of World Business*. 26(4):30–44.

Index

advantages of technological alliances: expected, 45–6, 64, 66, 86, 87, 104–5, 124; actually drawn, 45, 46, 57, 67, 87, 105, 124
Alberta, 34, 35, 57, 73
Aldrich, Howard E., 19
Allen, Robert C., 21
Arrow, Kenneth, 21, 25

British Columbia, 34, 72–3, 83, 97

Dasgupta, Partha, 26, 27
difficulties in technological alliances, 36, 46–7, 49, 57, 67, 87, 89, 91, 106, 124, 125
Dutton, John M., 21

European technological alliances, 12, 29–31, 39, 43, 49, 50, 61, 80
evolutionary theory and technological alliances, 15–22, 23, 128–9

Forester, Tom, 38, 60

Freeman, Christopher, 16, 20
Freeman, John, 19, 20n
Frischtak, C., 18

Hagedoorn, John, 9, 61, 75–6
Hannan, Michael T., 19
horizontal alliances, 4, 13–14, 100–1

industrial economics and technological alliances, 6–8, 22, 127
intellectual property, division of, xii, 3, 13, 44, 49, 54, 58, 66–7, 68, 69, 75, 85, 86, 104, 115, 117–20, 124, 125
international technological alliances, xii, 4, 14, 36, 39, 48–9, 50–1, 55–6, 61, 62, 64, 70–1, 72–3, 74, 92, 93–7, 108, 109, 111, 117, 123, 126

Japan, 35, 63, 101; and technological alliances, 27, 28–9, 36–7, 39, 61, 80

Korea, 101; and technological alliances, 50

legal arrangements for technological alliances, xii, 4, 13, 42, 49, 66, 115, 119, 124
Lundvall, Bengt-A., 10, 13, 20

management of technological alliances, xii, 43, 44, 54, 58, 64, 65, 69, 85, 86, 103, 108, 117–18, 124, 125
management theories and technological alliances, 9–11, 22
Manitoba, 34
Metcalfe, J.S., 15
Mowery, David, 9, 14, 51, 101, 102, 127

national technological alliances, xii, 4, 14, 50–1, 70, 73, 74, 92, 97, 104, 108, 109–11, 123, 124, 126
Nelson, Richard R., 5, 20

neoclassical theory and technological alliances, 4–6
New Brunswick, 34
Niosi, Jorge, 20
Nova Scotia, 34, 82
number of partners in technological alliances, 42, 43, 65, 85, 103

Ontario, 34, 35, 57, 70, 72, 82, 83, 96

partners, choice of, in technological alliances, 45, 53, 65, 85, 86, 104, 105, 124
patents in technological alliances, 13, 31, 44–5, 66, 79, 86–7, 94, 104, 118, 120, 123
Perez, C., 16
Porter, Michael, 11, 13
public financing of technological alliances, 25–37, 42, 50, 86, 104, 113–17; difficulties with, 47, 49, 69–70, 89, 91, 106, 115, 124, 125
public laboratories in technological alliances, 3, 4, 9, 11, 14, 25, 28, 33–4, 42, 51, 56, 57, 63–4,

70, 73, 75, 79, 82–3, 92, 108, 110, 113, 115, 116, 118

Quebec, 34, 35, 57, 72, 73–4, 82, 83, 95–6, 109–10

regional economics and technological alliances, 9
regional technological alliances, 50–1, 70, 71, 92, 93, 108, 109, 123, 126, 128–9
Rosenberg, Nathan, 5, 10, 14, 18, 51, 102, 127

Saskatchewan, 34, 82, 83
Saviotti, Paolo, 15
Schakenraad, J., 61, 75–6
Schumpeter, Joseph, 6–7
Simon, Herbert I., 15
Soete, Luc, 36
Stiglitz, Joseph, 17
Stoneman, P., 26, 27

technological alliances, definition of, 3
Thomas, A., 21
transaction costs and technological alliances, 8–9, 23, 98, 122, 127

United Kingdom, 26, 63, 77
United States, 26, 27, 36, 37, 53, 60, 63, 78, 79, 83, 92, 100, 101, 108, 111; technological alliances in, 28, 31–3, 39, 43, 49, 50, 55, 56, 61, 80
universities in technological alliances, 3, 4, 9, 11, 14, 24, 25, 31, 34, 39, 54–5, 56, 57, 64, 70, 72, 74, 75, 77–9, 83, 84, 92, 108; nonmarket approach, 47, 51, 128–9; and intellectual property, 116, 118, 120

vertical technological alliances, 4, 13–14, 100–1, 123
Von Hippel, 10, 13, 14

Watkins, Todd A., 27
Western Europe, 26, 28, 35, 36, 37, 56, 60, 63, 77, 100, 101. See also European technological alliances
Whitcomb, David, 22
Wilson, John, 63
Winter, Sidney, 5, 19, 46, 87, 123